EAST ASIAN CONFLICT ZONES

EAST ASIAN CONFLICT ZONES

Prospects for Regional Stability and Deescalation

EDITED BY
Lawrence E. Grinter
AND
Young Whan Kihl

ST. MARTIN'S PRESS
NEW YORK

© 1987 Lawrence E. Grinter and Young Whan Kihl

All rights reserved. For information, write:
Scholarly & Reference Division,
St. Martin's Press, Inc., 175 Fifth Avenue, New York, NY 10010

Printed in the United States of America

Library of Congress Cataloging-in-Publication Data

East Asian conflict zones.

    Bibliography: p.221
    Includes index.
    1. East Asia–Strategic aspects. I. Grinter,
Lawrence E. II. Kihl, Young W., 1932–
UA830.E25   1987   355'.03305   87-13118
ISBN 0-312-00377-3

To a safer world for this and
future generations.

CONTENTS

BIOGRAPHICAL SKETCHES
    ON THE CONTRIBUTORS   ix

PREFACE   xi

1. Conflict Patterns in East Asia and the Western Pacific   1
   Lawrence E. Grinter   Young Whan Kihl

2. Sino-Soviet Relations in the Late 1980s: An End to Estrangement?   29
   Steven I. Levine

3. Japan, the Soviet Union, and the Northern Territories: Prospects for Accommodation   47
   Peggy L. Falkenheim

4. Stability and Instability in the Sea of Japan   70
   Edward A. Olsen

5. The Korean Peninsula Conflict: Equilibrium or Deescalation?   97
   Young Whan Kihl

6. The South China Sea: From Zone of Conflict to Zone of Peace?   123
   Donald E. Weatherbee

7. Thai-Vietnamese Rivalry in the Indochina Conflict   149
   William S. Turley

8. Philippine Communism: The Continuing Threat
   and the Aquino Challenge   177
   Leif R. Rosenberger

9. Conclusion: Opportunities for
   Deescalating East Asia's Conflicts   206
   Young Whan Kihl   Lawrence E. Grinter

SELECT BIBLIOGRAPHY   221

INDEX   229

BIOGRAPHICAL SKETCHES ON THE CONTRIBUTORS

PEGGY L. FALKENHEIM teaches political science at the University of Western Ontario in London, Canada. She was Director of the University of Toronto's Office of International Cooperation. She has published articles on Soviet-Japanese relations that have appeared in *Asian Survey, Pacific Affairs, International Journal,* and several multiauthored volumes.

LAWRENCE E. GRINTER is Professor of Asian Studies, Air University, Maxwell Air Force Base, Alabama. A previous faculty member of the National War College and Air War College, he has published widely in his field including *Asian-Pacific Security* (1986, coeditor), and undertaken numerous studies for the National Security Council and the Office of the Secretary of Defense.

YOUNG WHAN KIHL is Professor of Political Science at Iowa State University in Ames. He was Visiting Scholar at the University of California, Berkeley, Institute of East Asian Studies in 1985 and also at the George Washington University Institute for Sino-Soviet Studies in 1978. A specialist in international and comparative (Asian) politics, his most recent books include *Asian-Pacific Security* (1986, coeditor), *World Trade Issues* (1985, coauthor), and *Politics and Policies in Divided Korea* (1984).

STEVEN I. LEVINE teaches East Asian politics at the American University in Washington, D.C., where he is Professor of International Service. He is the author of *Anvil of Victory: The Communist Revolution in Manchuria* (Columbia University Press, 1987), and has written extensively on contemporary Chinese foreign policy and related subjects.

EDWARD A. OLSEN is Professor of National Security Affairs at the Naval Postgraduate School, Monterey, California. A specialist in

Japanese and Korean affairs, he formerly served as a Northeast Asian Intelligence Officer with the U.S. Department of State. He is the author of numerous books and articles including *US-Japan Strategic Reciprocity* (1985) and *The Armed Forces in Contemporary Asian Societies* (1986).

LEIF ROSENBERGER is a Senior Analyst in the Strategic Studies Institute at the U.S. Army War College. He was a specialist in Soviet foreign affairs and counterterrorism with U.S. government agencies, and he also taught at Providence College, Rhode Island. His articles on international communism have appeared in *Problems of Communism*, *Survey*, and the Hoover Institution's *Yearbook on International Affairs*. He is the author of *Soviet Union and Vietnam: An Uneasy Alliance* (1986).

WILLIAM S. TURLEY is Professor of Political Science at Southern Illinois University in Carbondale. He was Fulbright-Hayes Visiting Professor at Chulalongkorn University, Bangkok, Thailand, from 1982 to 1984. He is the author of numerous studies on Vietnamese affairs including *Confrontation or Coexistence: The Future of ASEAN-Vietnam Relations* (1985, editor) and *The Second Indochina War 1954–1975* (1986).

DONALD E. WEATHERBEE is the Donald S. Russell Professor of Contemporary Foreign Policy at the University of South Carolina, Columbia. He also taught at other institutions including the University of Rhode Island, the U.S. Army War College, and was a Fulbright Fellow at the Institute of Southeast Asian Studies, Singapore, from 1981 to 1982. He has published widely on Southeast Asian affairs including *Southeast Asia Divided: The ASEAN Indochina Crisis* (1985).

PREFACE

East Asia and the Western Pacific is old in history and tradition, but it is also new in dynamic economic growth and conflict potential. In this sense, East Asia belongs to the future as much as to the past. The possibility of East Asia turning into the world's economic powerhouse is greater today than at any time in the past. Japan has surpassed the United States in per capita GNP and East Asia's Newly Industrializing Countries (NICs)—South Korea, Taiwan, Hong Kong, and Singapore—are performing remarkably well. In fact, East Asia is likely to emerge in the twenty-first century as the world's key economic region; the area's burgeoning economic growth and increasing linkage with North America and other regions are extraordinary.

However, East Asia and the Western Pacific is also one of the most heavily armed and dangerous regions of the world; some of its subregions and zones display lethal rivalries and competition. Accordingly, East Asia is in search of peace and stability commensurate with its economic performance. Thus, it is fitting that this present volume on East Asian conflict zones and the prospects for deescalation of tensions should be undertaken.

The major aims of this book are twofold: first, to clarify the dynamics and complexities of East Asia's most prominent conflicts and tension areas and second, to suggest what practical options, incentives, and means exist to deescalate and, if possible, to resolve these conflicts. For these purposes, the coeditors were fortunate in being able to assemble or draw on six leading scholars and analysts of East Asian security affairs to help carry out the project. Dr. Edward Olsen of the U.S. Naval Postgraduate School examined the Sea of Japan conflict zone; Dr. Peggy Falkenheim, a Canadian scholar, analyzed the Northern Territories' issue; Dr. Young W. Kihl of Iowa State University appraised tensions on the Korean Peninsula; Dr. Steven Levine of American University explored the Sino-Soviet dispute; Dr. William Turley of Southern Illinois University analyzed the Indochina conflict; Dr. Donald Weatherbee of the University of South Carolina evaluated the South China Sea zone of conflict; and Dr. Leif Rosenberger of the

Army's Strategic Studies Institute reviewed the Philippine situation. Dr. Grinter and Dr. Kihl jointly wrote the introductory and concluding chapters. Dr. Kihl and graduate assistants at Iowa State University compiled the Select Bibliography.

The origin of this particular book project was the two-tiered panel on East Asian Conflict Zones at the March 1986 International Studies Association Convention in Anaheim, California, in which most of the contributors presented their initial papers. The papers subsequently were revised and updated through 1986 and St. Martin's Press agreed to publish the book.

The coeditors wish to thank several individuals and institutions for their involvement in this project. First and foremost our thanks go to each contributor to this volume, for they represent some of the most active and respected analysts of East Asian security affairs in North America. The coeditors also wish to thank Mr. Kermit Hummel, Director, Scholarly and Reference Books, at St. Martin's Press, for his early interest in the project and his attendance at our panel at the Anaheim conference. Other members of the St. Martin's Press' editorial staff, Laura-Ann Robb and Amelie Littell, were also especially dedicated and skillful in their support.

Lawrence Grinter wishes to thank many of his faculty colleagues and students at Maxwell Air Force Base, Alabama, whose valuable discussion, comments, and support over the years have sharpened his appreciation of Asian-Pacific security problems. Young Whan Kihl wishes to thank Dr. Robert Scalapino, Director of the Institute of East Asian Studies at the University of California, Berkeley, for a productive stay as visiting scholar. He also acknowledges the receipt of an Iowa State University faculty improvement leave grant, in 1985, to initiate a study of East Asia's regional security and political economy. Three graduate students assisted Dr. Kihl on the Select Bibliography: Jeffery Beattie, Victor Foggie, and Sung-gol Hong.

The coeditors and contributors hope that this book will make a substantial contribution, on both sides of the Pacific, to a better understanding of the potentialities for reducing tensions and bringing about peace in East Asia and the Western Pacific.

EAST ASIAN CONFLICT ZONES

1. Conflict Patterns in East Asia and the Western Pacific

Lawrence E. Grinter and Young Whan Kihl

EAST ASIA is today one of the most economically dynamic regions in the world, the region which provides both a competitive edge and an economic challenge to the West. East Asia is becoming the world's most productive region whose goods, technology, and services are outcompeting the West in many instances.[1] East Asia, however, is also a conflict-ridden region, as subsequent discussion in this book will amply demonstrate. It is a region heavily armed militarily, a region where the interests of four major world powers—the United States, the Soviet Union, China, and Japan—converge and crisscross. The nuclear buildup in the Soviet Far East has accelerated in the recent decade, thereby posing serious policy dilemmas for the United States and its allies in Asia.[2]

This introductory chapter examines some of the ongoing military, territorial, political, and ethnic conflicts in East Asia by placing them within the context of the most significant interactions and rivalries in the area—global, regional, and local. The present chapter will also suggest the range of practical policy options for the United States and its allies in managing and attempting to direct the conflict patterns toward deescalation and stability.

I. An Overview of the Region and Conflict Issues

For the purpose of the present discussion, we define East Asia and the Western Pacific as that portion of Asia that consists of three main subareas—Northeast Asia, Southeast Asia, and Pacific Oceania. The countries of Northeast Asia are Japan, China, North and South Korea, Mongolia, and Taiwan. Southeast Asian countries include Burma, the three Indochinese states of Vietnam, Laos, and Kampuchea, and the six ASEAN (Association of South East Asian Nations) states of Thailand, Malaysia, Singapore, Indonesia, Brunei, and the Philippines. In Pacific Oceania, Papua New Guinea, Australia, and New Zealand are included.* The United States and the Soviet Union, which have major interests, forces, and/or treaties in each of these subareas, naturally are also included in the present analysis.

Given the importance of the region, the major powers have been strengthening their security postures in the area in the 1980s. For example, both strategic and conventional forces, all undergoing modernization, are deployed in the region by the Soviet Union, China, and the United States. The most tense areas of the region have pitted millions of hostile troops against each other: along the Sino-Soviet border, across the Korean DMZ, at the Sino-Vietnamese border, and within Indochina. The Soviet Pacific Fleet at Vladivostok is Moscow's largest navy, and about one third of Soviet SS-20 IRBMs (Intermediate Range Ballistic Missiles) are deployed in Asia. In Northeast Asia, the Russian fleet and the U.S. Seventh Fleet face each other across the Sea of Japan. The two antagonists' navy and air force elements also confront each other across the South China Sea in Southeast Asia; elements of Moscow's Pacific forces operate out of Cam Ranh Bay and Da Nang in Vietnam, while U.S. forces are positioned at Subic Bay and Clark Air Base in the Philippines. As Soviet force deployments have increased in the area and, in essence, encircled China, ASEAN countries have reacted by increasing their own defense expenditures and shifting their military doctrine away from counterinsurgency to external defense. At the subregional and local levels, a host of territorial and ethnic disputes also complicate the region's stability: on the Sino-Soviet border, in the

---

*Excluded from the scope of this chapter are certain micro-states in the Pacific such as those in Micronesia (TTPI—Trust Territory of the Pacific Islands), Western Samoa, Tonga, Fiji, and other U.S., French, and British dependencies. Hong Kong, a British dependency, reverts to China in 1997.

southern Kurile Islands between Japan and the USSR (Union of Soviet Socialist Republics), throughout Indochina, between Vietnam and Thailand, and in areas of the South China Sea where ASEAN, Vietnamese, Chinese, and Taiwanese interests all come into play. Finally, there are the ongoing guerrilla wars in the Philippines.

The Region's Economic Dynamism

Conflicts in East Asia and the Western Pacific will thus affect the region's and the world's stability because they tend to jeopardize the region's extraordinary trade and investment growth. Led by the "Gang of Five" (Japan, South Korea, Taiwan, Hong Kong, and Singapore), Asia-Pacific has been achieving 8 to 10 percent annual growth rates even when the poor performance of the Indochinese states and the Philippines, and some slowdowns by other countries, are included. When North American (especially U.S.) trade with the region is added in, and the Asian-Pacific region is broadened to include the entire Pacific Basin, the region also is the "high-technology" center of the world, bounded by Silicon Valley on the U.S. West Coast, by Japan, Taiwan, and Hong Kong in Northeast Asia, and by Singapore in Southeast Asia. Total U.S. two-way trade with East Asia and the Western Pacific in 1985 was estimated to have been in the vicinity of $180 billion—greater again, for about the seventh year, than U.S. trade with the European Common Market and almost one third of all U.S. trade.[3] For 1986, total U.S. trade with East Asia topped $190 billion and is projected to be over $200 billion for 1987. By comparison the Soviet Union probably does less than $20 billion of trade with the area.[4]

Certain Asian-Pacific countries have become economic heavyweights. Japan's GNP in 1985 was over $1.5 trillion, the second largest in the world, and Japan and the United States together accounted for about one third of all global economic productivity, with their combined share of world trade approximately 25 percent. South Korea's economic growth also continues to be impressive, with its $85 billion GNP in 1985 about four times as large as North Korea's. In Southeast Asia, with the exception of the Philippines, the ASEAN countries generally continue to show excellent progress. Between 1973 and 1983 ASEAN countries averaged annual growth rates of 7.6 percent.[5] Their combined exports in 1984 were over $70 billion and their per capita income growth also has been impressive. For example, compared to Vietnam's meager $180 annual per capita income, Thailand's in 1984 was about $800. Singapore's in 1984 was over $5,000, and oil-rich Brunei's was pushing $20,000. ASEAN's combined population was almost six times Vietnam's, and ASEAN's combined GNP, at over $200 billion, was fifteen times larger than Vietnam's. As a group, ASEAN now forms the United States' fifth largest trade partner.[6] In short, East Asia's economic progress, backed by U.S. trade with and investment in the area, has propelled the region into the position of the world's emerging economic leader—the basis of a global "Pacific-Asian Era." These economic stakes make the region's conflicts all the more critical.

The Geopolitical Role of East Asia

U.S.-Soviet relations in East Asia and the Western Pacific reflect the imperatives of the two countries' global rivalry. Their respective security policies are heavily dominated by geopolitical considerations and resource endowment. Since the USSR, as a Eurasian land power, is pursuing a great power geopolitical role in the region, it alternates border security operations, expansionism, and recently, under Gorbachev, new attempts to complicate relations among the United States, its allies and friends. The United States, as a global maritime power with regional interests in East Asia, but with diminished military capacities and economic power in the area, is pursuing policies to counterbalance the Soviet and Soviet client expansionism in Asia by allying itself with the regional powers facing or acknowledging the Soviet threat. Whereas the Soviet Union operates from its historical role of a land-based continental state, the United States must continue to rely on a maritime, coalition strategy to offset, perhaps to encircle, the Soviet Union.

Halford Mackinder's celebrated "Geographical Pivot of History" thesis and Alfred Mahan's "Sea Power Domination" hypothesis, or what one student appropriately called "a marriage of Mahan and Mackinder," seems to describe the dynamics of the U.S.-Soviet rivalry in Asia and the Western Pacific in the 1980s. This divergence in the superpowers' respective roles and strategies is poignantly depicted by an Australian observer in a somewhat pessimistic mode:

> ... to be the eventual predominant power of Euro-Asia, Moscow doesn't have to do anything more than it has done for several centuries: press outwards from the pivot (set at about the Ural mountains) operating from internal lines of communication, and acting out the imperatives of power and the corruption of opportunity. For the US to exercise constraints on Soviet power in Asia, it cannot be other than a political, cultural and geographical intruder operating over long distances on external lines of communication at the margins of an essentially alien continent. This applies to US use of nuclear as well as conventional forces.[7]

Fortunately for the United States, the strategy of maritime and rimland coalition politics against Soviet and Soviet client expansionism in Asia has been responded to positively by its allies and friends in the region. Japan is interdependent with the United States and Australia in playing a positive role in the global balance. Japan serves as the principal U.S. forward base for the containment of Soviet power in the Northwest Pacific and Northeast Asia. China, with its own reasons for opposing the USSR, has become the eastern counterweight to the USSR, "the giant at the back gate."[8] ASEAN serves as an emerging if still modest bulwark to Vietnamese intrusion into maritime Southeast Asia. The strategic importance of Northeast Asia in the global balance is self-evident, as the following description shows, and its strategic value may be ignored at greater risk and considerable opportunity costs.

> Northeast Asia is an area of dangers to world peace because it provides the nexus between four great powers with competing ambitions: the Soviet Union, determined to develop the resources of Siberia and to have unimpeded access to the Pacific for mercantile shipping and the projection of naval power; China, determined to be influential over its continental sphere; Japan, a maritime power, lying across the Soviet exits and dependent on the US for protection against Soviet hegemony; and the US, dependent on Japan for its Western Pacific strategic presence. The Korean peninsula lies at the nexus, manifesting by its division the competing ambitions, pulled and pressed within and without, a self-propelled pawn in a complex power game.[9]

In 1986 the Soviet Union, not surprisingly, proclaimed itself an "Asia-Pacific country." During his July 28, 1986, Vladivostok speech on the "Soviet Role in Asia," to be discussed below in detail, Soviet leader Mikhail S. Gorbachev called for further economic development of the Soviet Far East, while asking at the same time for better relations with China and Japan, as well as for the calling of a future Pacific Security meeting.

II. The Soviet Problem

The most serious single source of tension in East Asia and the Western Pacific is the expansionist policies of the Soviet Union. The USSR, a Eurasian country, views Asia-Pacific as an insecure frontier zone toward which it must erect a security belt of docile or friendly states, linked to the Soviet Union by military infrastructure and/or political-economic dependencies. The USSR shares the longest border in the world with its most critical adversary in the region: the Sino-Soviet border of 4,000 miles. Since 1965, when the Soviets began their determined military buildup in the region using ground, air, and naval assets, they have encircled China with the largest quantities of modern combat power in East Asia: on the Sino-Soviet border fifty-two Soviet divisions are augmented with first-generation tactical and strategic aircraft and missiles; at Vladivostok, on the Sea of Japan, Russian naval tonnage outnumbers the U.S. Seventh Fleet five to two; at Cam Ranh Bay in Vietnam, twenty to twenty-six Soviet ships and up to six submarines operate 800 miles across the South China Sea from U.S. assets at the Philippines' Subic Bay and Clark Air Base; and in Afghanistan, 110,000 Soviet troops wage war on the Afghan rebels.[10] All of these operations depend principally on military power, the Soviets having made little effort until Secretary General Gorbachev's Vladivostok speech to balance their military efforts with economic and political measures or proposals.

The Soviet Union's military buildup in the Far East is augmented by deployment of nuclear weapons. Since the late 1970s, Soviet nuclear delivery systems directed against potential Asian targets have undergone a qualitative change, with the deployment to the Far East of Badger and Backfire bombers and the SS-20 IRBM. Approximately 80 to 120 bombers, or about one-third of the Soviet inventory, are said to be deployed in Asia, augmented by their capability to carry the AS-4 air-to-surface missile.[11] The SS-20 missiles, with maximum range of 5,000 kilometers (3,125 miles), are now MIRVed (Multiple Independently Targeted Reentry Vehicle) with three warheads each and also equipped with mobile launchers which are difficult to detect. Approximately 135 to 171 SS-20s are said to be deployed in the Far East theater east of the Urals.[12]

In spite of the Soviets' power projection policy in Asia and the Pacific, which in part relies on their basing out of Vladivostok and Cam Ranh Bay, and their support of state-sponsored terrorism and military operations by North Korea and Vietnam, there have been indications that the Soviets have not been particularly satisfied with their overall prospects in Asia and the Pacific. Certainly the July 1986 proposals by Gorbachev, and the activities of his foreign minister, Shevardnadze, suggest a long-overdue effort by Moscow to change its image in East Asia. First of all the Soviets have not been able fundamentally to recompose their relations with China. To the west in Afghanistan, the Soviets have admitted fatigue with their protracted Afghan operation, now into its eighth year. To the south the Soviet's client, Vietnam, has yet to pacify Kampuchea, or bring world opinion to its side, and it is costing Moscow at least $2 billion per year to sustain Vietnam and Hanoi's client in Kampuchea. And North Korea's periodic terrorism against South Korea has put both the USSR and the DPRK (Democratic People's Republic of Korea—North Korea) on the defensive in Northeast Asia. Regarding territorial issues, the Russians have not yet shown any real willingness to make the territorial or political compromises necessary fundamentally to improve relations with either Japan or China. Moreover, in comparison to the ASEAN countries and South Korea, the Soviets' associates in Hanoi and Pyongyang are essentially frozen out of the region's burgeoning economic growth. In short, the earlier policies of the Gromyko period perpetuated a frozen, hard-line situation which has won Moscow, Hanoi, and Pyongyang no new friends in East Asia and the Western Pacific. Gorbachev and Shevardnadze are clearly attempting to change this.

Until late in 1985, Mikhail S. Gorbachev, the fourth Soviet leader in three and one-half years, was preoccupied with his internal power consolidation, European affairs, and Soviet-American relations. Then, in early 1986, came the first indication that Moscow was seeking to improve its relations with East Asia, and (not surprisingly) at American expense: in mid-January Foreign Minister Eduard Shevardnadze made a five-day trip to Japan, where he pressed the Japanese to separate themselves from certain U.S. policies, including participation in the Strategic Defense Initiative (SDI). As the Soviets work to improve their diplomatic and political position in East Asia and the Western Pacific, opportunities certainly exist for them to do so: on the Sino-Soviet border, in Afghanistan, in the southern Kurile Islands as they affect Soviet-Japanese relations, and in Indochina. Of course, these opportunities existed before Gorbachev came to power in March 1985, but his predecessors showed no inclination to do anything about them. But since Gorbachev has been promoting detente in the West, we can safely assume that he may try something similar in the East. Moscow's long-term goal of breaking up the de facto U.S.-Japanese-Chinese-ASEAN alliance may be more successful if Gorbachev and Shevardnadze demonstrate more sophistication and dexterity than their predecessors whose hard line, if anything, drove these countries closer together.

Gorbachev's New Initiative on Asia

The new Soviet initiative on Asia finally came about in the summer of 1986. Soviet leader Mikhail S. Gorbachev delivered a major foreign policy address in Vladivostok on July 28, 1986, containing references to a wide range of domestic and foreign policy topics, including the Soviet Union's future relationship with the Asian-Pacific region.[13] The major Asian components and highlights of his address were:

- Soviet troop reductions in Afghanistan;
- Improving Soviet-Chinese relations;
- Calling for an Asian-Pacific security conference.

Clearly aware of previous Soviet failures to place themselves on the crest of Asian-Pacific events, Gorbachev proclaimed: "We are in favor of building new and equitable relations with Asia and the Pacific." After expressing the Soviet desire for improving ties with some Asian countries, including Japan, Gorbachev stated:

"By the way, with time, we might solve the question of opening Vladivostok to visits by foreigners; we would like it to be our widely opened window to the east." As Petersburg (the old Leningrad) was called by Czarist Russia "the window to the West" in the nineteenth century, so Vladivostok—the "fortress of the East" which is forbidden to foreigners thus far—will become the Soviet "window to the East" in the twenty-first century.

On Afghanistan, Mr. Gorbachev said that six regiments would be withdrawn before the end of 1986—one armored regiment, two motorized rifle regiments, and three antiaircraft artillery regiments—with their integral equipment and armaments "in such a way that anyone interested can easily verify it." He insisted, however, that the Soviet withdrawal "must be answered" by a reciprocal curtailment of Western aid to the guerrillas fighting Soviet and Afghan government forces.

On China, Mr. Gorbachev proposed talks on reductions of Soviet and Chinese land forces as well as the best ways of achieving the two countries' respective "priorities in developing and modernizing their economies." He claimed that "a noticeable improvement has occurred in our relations in recent years. I would like to affirm that the Soviet Union is prepared—any time, at any level—to discuss with China questions of additional measures for creating an atmosphere of good-neighborliness. We hope that the border dividing—I would prefer to say, linking—us will soon become a line of peace and friendship."[14]

To realize these aims, the Soviet leader mentioned several specific measures. First, Gorbachev disclosed that the Soviet Union and China were working out an accord to renew joint water management projects in the Amur River Basin on their eastern border, the project initiated but soon suspended in the 1950s. Second, Gorbachev said that the Soviet Union was preparing a "positive reply" to a Chinese proposal to renew construction of a railroad linking the Xinjiang Uighur Autonomous Region of China and the Soviet republic of Kazakhstan, a project also initiated and soon suspended in the 1950s. Third, the Soviet leader added that Moscow had offered to train Chinese astronauts for a joint space mission.

Most important of all was Moscow's offer to negotiate the Soviet troop reduction along the Soviet-Chinese border, including the withdrawal of Soviet troops from Mongolia. Gorbachev's offer to remove Soviet troops from Mongolia, if implemented, seems comparable to President Nixon's removal of U.S. troops from Taiwan and elements of the U.S. Seventh Fleet deployed in the Taiwan Straits on the eve of Nixon's China trip in February 1972. Gorbachev's proposals represent a "peace offensive" foreign policy, an evidently serious departure from Brezhnev's and Gromyko's overly militarized and hard-line foreign policies in Asia.

Regarding other Asian countries, Gorbachev called for an improvement of relations between the Communist countries of Indochina—Vietnam, Cambodia, and Laos—and the ASEAN countries of Indonesia, Malaysia, Brunei, the Philippines, Singapore, and Thailand. He underscored Moscow's appeal for "joint ventures" with Japan and ASEAN countries. Without conceding substance on Japan's Northern Territories claim, Gorbachev is eager to open the road for greater economic cooperation with Japan and also to discourage the latter from cooperating with the United States on SDI plans. The prospect of a mutual exchange of visits at the highest level, including Gorbachev-Nakasone summitry, is in the offing. If realized, this will be the first head-of-state diplomatic visit since 1972, the year Japanese Prime Minister Kakuei Tanaka made an official visit to Moscow without much tangible result.

On Asia as a region, Gorbachev proposed a regional conference on Asian-Pacific security which will be "attended by all countries gravitating toward the Pacific Ocean," including the United States. As the conference site, he offered one of the Soviet maritime cities, such as Vladivostok, although he also mentioned Hiroshima as a possible city to convene a Pacific Ocean disarmament conference. The Asian security conference, like the 1975 Helsinki Accord in Europe, would promote "the practical discussion of confidence-building measures and the nonuse of force in the region." Gorbachev seems determined to repair the damage to Soviet diplomacy in Asia perpetuated by Brezhnev's and Gromyko's hard-line militaristic policies.

Regarding relations with the United States in Asia, Gorbachev also proposed Soviet-American negotiations aimed at a reduction of Pacific naval forces, particularly nuclear-armed ships, and limitations on the deployment of antisubmarine weapons. He suggested that the United States consider removing some or all of its military forces from the Philippines. "If the United States were to give up its military presence in the Philippines, let's say, we would not leave this step unanswered," he said.

Countermeasures Strategy

It is reasonable therefore that the new initiative of Gorbachev and Shevardnadze represents a determined attempt to work the USSR out of its unsatisfactory position in East Asia. Such a Soviet "break out" or "comeback" strategy[15] likely has as its goals the weakening of the U.S.-Japanese-Chinese-ASEAN alliance and the enhancement of Soviet influence. In Northeast Asia, Moscow has the opportunity to try to complicate Japan's and China's links with the United States. The Russians may seek to work upon Japanese pacifism by proposing some kind of a neutralization agreement for the southern Kuriles. This could perhaps be accompanied by Soviet force reduction proposals near the Japanese home islands in return for weakened Japanese support for U.S. policies or military cooperation. Toward China, the Russians could meet Beijing's force pullback conditions, either partially or fully, in return for increasingly better relations with Beijing and an inevitable complication of Sino-American relations. In Indochina, the precedent exists for a major Soviet diplomatic initiative—one recalls the 1962 U.S.-USSR-U.K. accord, when the Soviets co-sponsored a "neutralization" of Laos which later formalized into a tripartite government arrangement in Vientiane. It may be that the Soviets will offer a force separation arrangement in Kampuchea between the Heng Samrin government and the Sihanouk resistance coalition, while also offering their good offices to both Vietnam and Thailand on the flanks. Such a Soviet offer and initiative would align Moscow with the interests of peace while also enlarging the Russians' influence in mainland Southeast Asia. Should Thailand and the Sihanouk coalition agree, anxiety would occur within other member governments of ASEAN, and U.S. diplomacy in the area could be thrown off balance.

This brings us to the question of U.S. policies and the policies of our principal friends in Asia and the Pacific toward the Soviets. Essentially the United States and its associates have three policy choices: (1) seek to further deter and ostracize the Soviets and their clients from a "breakout" strategy; (2) seek to give the Soviets and/or their clients a stake in the emerging prosperity of East Asia; or (3) use a combination of (1) and (2)—a kind of mixed deterrence plus detente strategy. Fundamentally, Soviet policy in East Asia still remains heavily influenced by domestic constraints, such as the need to revitalize the sagging economy, as well as the strategic problem with China.[16] Soviet relations with North Korea and Vietnam are affected by the difficulties between Moscow and Beijing. Of course, Soviet-DPRK and Soviet-SRV relations have their own origins and sensitivities. But Moscow's problems with Beijing drive the Soviets to press for separate and manipulable relations with Pyongyang and Hanoi. The Chinese reciprocate. China has, from the position of its own self-interest, chosen to conduct a policy toward the Soviets of mixed carrots and sticks. Beijing offers Moscow the prospects of a normalization in return for conditions which, if the Soviets met them, would clearly lessen tensions and promote stability in East Asia. China's threefold conditions, the last of which Deng Xiaoping recently called the "prime obstacle," are:

(1) Soviet force pullbacks on the Chinese border;
(2) a Soviet withdrawal from Afghanistan; and
(3) Soviet withdrawal of support from Vietnam's occupation of Kampuchea.

Thus the Chinese, who have dealt with the Russians longer than any other East Asian country, have chosen an approach toward Moscow which incorporates both reconciliation and firmness. This approach may also help the United States in developing a framework for overall U.S. policy toward the Soviets in Asia and the Pacific, and one that can, thereby, help to lessen conflicts in the region as a whole.

Although Beijing's initial response to Gorbachev's July 1986 Vladivostok statements regarding Afghanistan and Mongolia was "cool" and pro forma, reiterating China's position on "the prompt and complete withdrawal of Soviet troops," China's subsequent reactions were more "positive" and accommodating.

China's top leader, Deng Xiaoping, in an interview with CBS's "60 Minutes," offered to go to Moscow, a trip he has repeatedly refused to make thus far. Deng said, "If Mr. Gorbachev removes the three big obstacles, especially persuading Vietnam to stop the invasion of Cambodia and to withdraw its troops, then I myself would like to meet him."[17] Soviet leader Gorbachev had offered to meet with Deng several times in the preceding six months, but Deng had reportedly refused.

China's Deng, in a clever move, thus sought to steal the diplomatic limelight from Gorbachev. The diplomatic ball was back in the Soviet court, both sides knowing full well that the Soviets' clout on Vietnam's Cambodian occupation is limited and "sensitive." Since Vietnam claimed to have withdrawn some of its troops already from Cambodia, and also proposed withdrawing all of them by 1990, the Vietnam issue may not pose a formidable obstacle to realizing a Deng-Gorbachev summit. In fact, Deng did not ask the Soviets to bring about the *complete* withdrawal of Vietnamese troops from Cambodia and some substantial reduction of troops might easily be acceptable to him.[18]

III. The Emerging Asian-Pacific Alliance System

Within the context of the superpower competition in East Asia, and the extraordinary economic vitality of the region led by the capitalist countries, a diverse group of East Asian and Western Pacific nations have drawn together into a relatively cohesive body in recent years. This association has been prompted by the convergence of common economic interests and practices, and by mutual reactions to the security threats emanating from the Soviets' and their allies' activities in the region. The emerging Asian-Pacific coalition is led by the United States and consists of the following: the United States, Japan, the People's Republic of China (PRC), South Korea, the ASEAN countries, and Australia. The alliance policies of each of these key members will now be examined.

The United States

In the last six years the Reagan administration has refocused U.S. policy priorities in Asia and the Pacific on five central pillars:[19]

—The continuing, key importance of U.S.-Japanese relations;
—U.S. efforts to build an enduring relationship with China;
—Maintaining stability on the Korean peninsula;
—Supporting ASEAN;
—Maintaining support for ANZUS, particularly Australia.

These five relationships are the foundations of United States policy in the region. With less forces in East Asia and the Western Pacific than the other major powers,[20] and the U.S. West Coast being thirty naval steaming days from the region, the United States finds itself increasingly dependent on other countries in Asia and the Pacific to maintain the necessary strength to deter the Soviets and their clients.

At the strategic level, the combined effects of U.S. air and naval strategic power in East Asia, backed by U.S. ground troops in Korea and offshore, continue as the single most critical deterrent to Soviet expansion on Asia's rimlands. On the continent, China's large ground forces, and their territorial bulge into the USSR's eastern holdings, constitute the major deterrents to Soviet outreach. On the peripheries, South Korea's well-armed and highly disciplined armed forces, and the Kampuchean guerrillas backed by Thailand and ASEAN, constitute flanks of a second buffer zone.

China

Although Chinese spokesmen go to considerable lengths to deny that there is a strategic partnership between the PRC and the United States,[21] and until well into the Reagan administration's first term the White House was uncomfortable with the idea,[22] it is obvious that from the perspective of their shared common security interests and their growing economic and technical/military trade the United States and the People's Republic of China are de facto security allies with the common objective of deterring the Soviet Union and complicating its aggressive outreach.[23] Mao Zedong is reputed to have once stated: "What is detente? I am detente. Without me Russian divisions would have overrun Western Europe long ago." There is certainly truth in this view that China, by its massive bulk, ground forces, and racial/ideological/territorial opposition to the Soviets, acts as a critical deterring factor to a Soviet-dominated East Asia and the Pacific. Should the Soviets and the Chinese resume their alliance, thus allowing military power to be deployed out of Chinese air and naval bases, the political, and most likely the geographic, map of East Asia would dramatically change.

Japan

As the United States' principal ally in East Asia, the Japanese, in spite of their relatively low defense expenditures, remain the keystone to allied security in the region. Ideally located for augmenting U.S. military protection against the Soviet and North Korean threats in Northeast Asia, Japan, in the words of Prime Minister Nakasone, has acted as an American "aircraft carrier." The problems in the U.S.-Japanese security relationship, however, are real, and derive in the American view from Japan's continuing low level of defense spending, coupled with its unwillingness to defend adequately its own territory and sea straits against Soviet military pressure. Without U.S. forces, Japan would be unable to stand up to Soviet power. This is, of course, a deliberate and still popular policy by the Japanese[24] who do not—after having fought the Russians three times in this century—care to match them in or threaten them with military power.

U.S. officials have been candid about Japan's minimal defense effort. For example, in November 1982, Admiral Robert Long, Commander-in-Chief Pacific, stated: "The Japanese are individually well-trained, well-disciplined and technically very competent. The major problem is that they lack adequate supplies of fuel, ammunition and missiles. In my judgment, they lack the ability to handle even a minor contingency."[25]

Japan's recent Mid-Term Defense Program (1983–87) fell short of its goals.[26] While addressing critical current problems like replacing obsolescent equipment, improving training, increasing ammunition supplies, expanding transportable capability, and developing an integrated strategy and concept for operations, it is doubtful that the deficiencies pointed out by U.S. officials will be overcome by the early 1990s. And still Soviet capabilities grow. The Japanese Defense Agency's (JDA) own analysis is sobering.[27]

South Korea

Well armed and competently led, South Korea's armed forces are having to protect a country whose capital is just 30 miles from North Korean forces on the DMZ. North Korean missiles and guns are poised against the Republic of Korea (ROK), and Seoul is just three minutes' air time from North Korean–piloted Soviet jet aircraft. Facing North Korea's armed forces of 784,000 men, including 100,000 commandos, South Korea is also threatened by 3,400 tanks, 4,600 to 5,000 artillery guns and howitzers, perhaps three fourths as many as the U.S. Army has worldwide.[28] Russian-made Frog 5 and Frog 7 missiles close to the DMZ could hit Seoul in a matter of seconds. From the October 1983 Rangoon bombing, to continued tunnel digging and infiltration attempts, evidence of DPRK hostility to the ROK continues.

Against this unrelenting threat, the Seoul government's familiar problem continues: how to maintain a sufficient deterrent capability, military and psychological, while also incorporating the demands of a rising middle class and articulate political oppositionists like Kim Dae-Jung and Kim Young-Sam. Neither man has been able seriously to challenge President Chun's power so far despite sharp parliamentary challenges and student activism which has put the Chun government off balance.[29] South Korea's international image has benefited from recent and upcoming athletic-diplomatic events, like the 1986 Asian Games and the 1988 Seoul Summer Olympics. President Chun, whose own prestige should benefit from these events, has promised not to run for reelection in 1988.

ASEAN

Having emerged in the last few years as a remarkably successful economic and diplomatic entity, the Association of Southeast Asian Nations faces the late 1980s confronting a changing external security environment, and critical internal political transitions within some member states. Noting the enlargement of Soviet forces based out of Vietnam's facilities, and Vietnam's combat operations in Kampuchea, ASEAN's armed forces have been shifting from an internal counterinsurgency focus to conventional capabilities oriented on the external threat. Large increases in ASEAN defense budgets have occurred, as well as increasing standardization of equipment and training, and there have been more multilateral military exercises.[30]

However, as an alliance of developing countries, ASEAN's members also show continuing difficulties in their internal patterns of political power sharing and economic development. In Thailand, in September 1985, another attempted coup d'état fizzled out after some loss of life. In Indonesia and Singapore, political successions may be under way as President Suharto and Prime Minister Lee confront the need to prepare leadership turnovers. And in the Philippines, Corazon Aquino's succession to the presidency after Ferdinand Marcos fled the country has produced a number of valuable initiatives, but the country's fundamental problems remain very severe.

Australia

Within the ANZUS alliance, the bilateral security arrangements between Australia and the United States have survived, possibly even been somewhat strengthened by New Zealand's break-away on the nuclear issue. By its insistence on no nuclear-powered or -armed ship visits, by any countries, the Lange government in Wellington ended up terminating its military cooperation with the United States. As a result, Australia's importance to regional security affairs has increased and, with it, so has United States dependence on Australia.

These six actors then—the United States, Japan, the PRC, South Korea, ASEAN, and Australia—constitute the new Asian-Pacific alliance system. The problem for them is to manage the shifting threats, challenges, and conflict patterns in the late 1980s so that coalition policies can be developed which will (1) maintain the convergence of the partners' shared goals; (2) present a cohesive deterrent strategy to the Soviets and their clients; and (3) solve coalition problems for the good of the whole.

IV. Building an Allied Strategy of Stability and Deescalation in East Asia: "Deterrence Plus Detente"

The key to East Asia and the Pacific's long-term security stability will involve a strategy of dealing with the Soviets and their allies so as to temper their behavior while also finding solutions to local problems within our coalition. Toward the Soviets, an allied coalition policy similar to what the Chinese offer Moscow makes sense: holding out benefits in return for Soviet force deescalation and policy moderation. If the Soviets can be netted into a broader system of cooperative behavior—i.e., deterrence plus detente—and their often friction-prone relations with North Korea and Vietnam exploited, this will help to shield other problems in the region where a Soviet exacerbation has to be discouraged—particularly in Kampuchea and the Philippines, but also in the South Pacific.

Dealing with the Core: The Soviets

Given that Moscow still does not enjoy much trust or political influence in East Asia and the Pacific, the United States and its partners can exploit these facts with a series of economic-political inducements to the USSR to reconsider its policies. Indeed, elements of the Gorbachev proposals at Vladivostok ought to be tested and exploited. Russian force deployments around China's border, and the shrinking capacity of its Siberian oil industries,[31] have produced serious long-term complications for the USSR's economic strength and projection into the region. China's patient policy of offering better relations to Moscow in return for Soviet force disengagements makes sense and probably influenced Gorbachev's offer of concessions toward China. In addition, the periodic discussions between Tokyo and Moscow on exploitation of Siberian energy reserves could be rekindled. In return for Soviet force reductions in the Sea of Japan, and an end to Russian air and naval violations of Japanese territorial space, Tokyo could offer new negotiations on economic exploitation of the Soviet Far East. These Chinese and Japanese initiatives could become part of a broader allied diplomatic agenda presented to Moscow and covering major conflict zones on the Sino-Soviet border, in Afghanistan, in Indochina, in the Sea of Japan, and in the South China Sea.

There are, of course, risks to U.S. interests in supporting such far-ranging negotiations, including separate negotiations between Beijing and Moscow, and between Tokyo and Moscow. Should the Russians take either country up on their proposals, and a warming of relations between either and Moscow occur, further complications may arise in relations between Beijing, Washington, and Tokyo. Ultimately the United States and its major partners in East Asia, however, will need the sense of mutual confidence that "separate deals" will not separate the allies, or be made at the expense of a third party.

Dealing with the Peripheries

While encouraging Soviet force reductions along China's borders, in the Sea of Japan, in Indochina, and in the South China Sea by supporting Chinese, Japanese, and ASEAN initiatives, U.S.-led policy also should concentrate on exploiting North Korea's and Vietnam's problems. The principal "front-line" states here are, of course, South Korea and Thailand. Long-term allied policies of mixed deterrence and detente toward Pyongyang and Hanoi should prove fruitful. It will be a process that could last decades. Both the DPRK and the SRV need better economic prospects, both are wasting enormous resources on abject military ventures, both suffer from tarnished reputations and general opposition in East Asia and the Pacific, and both are confronted by neighbors which are sharing in and shaping East Asia's economic boom. Finally, both North Korea and Vietnam are undergoing, or soon will undergo, major leadership transitions. With the emergence of Kim Jong Il in North Korea and Nguyen Van Linh in Vietnam, opportunities exist for new allied initiatives aimed at enticing both regimes into more moderate behavior in return for economic-technical rewards and a relaxation of tension. Such a result would have two advantages for the U.S.-led coalition: more responsible and restrained behavior by Pyongyang and Hanoi would help stability in Northeast Asia and Southeast Asia, and could also further complicate the ties between the DPRK, the SRV, and the USSR.

Toward North Korea. The sheer record of rapid economic growth in South Korea has had a sobering effect on Pyongyang: The ROK's vibrant economy, benefiting from an export-led development strategy, continues to push Seoul away from Pyongyang in their economic competition. With 42 million people and a 1986 GNP of about $90 billion, the ROK contrasts sharply with the DPRK's 21 million people and estimated $25 billion GNP.[32] At the most, North Korea's per capita income is no more than one half of South Korea's. It is not known to what extent the ROK's example, or the bottlenecks within North Korea's command economy, drive up the frustration levels within the DPRK's leadership. But terrorist outrages such as the October 1983 Rangoon bombing, and continued DPRK commando/subversive actions against the ROK, reflect a continuing desire by Pyongyang to try to stop South Korea's progress and decapitate its government.

From a position of allied diplomatic solidarity, then, South Korea and the United States (with implicit Japanese and Chinese backing) could beneficially propose to North Korea that it sign a permanent peace treaty with South Korea in return for its access to more trade, technology, and investment. Indeed in late 1984, Pyongyang enacted a limited foreign investment (i.e., joint venture) law, modeled after those trends in the PRC, which seeks to attract foreign investment.[33] Pyongyang will have to understand, however, that it cannot require a U.S. military withdrawal from South Korea as a *precondition* for a peace treaty. As treaty allies, one may argue that American forces stationed in South Korea are a matter solely between Seoul and Washington. (Just as Soviet and Chinese military relations with North Korea are matters solely between these three allies.)

But, if serious discussion toward a peace treaty should commence, other issues could be entertained which, if resolved, would clearly contribute to deescalation and tension reduction on the Korean peninsula and in Northeast Asia. These include conventional arms reduction. There is little sense in the two Koreas continuing to pile up arms along what is already probably the most heavily armed 155-mile strip in the world. North Korea must be persuaded that it is actually in its interest to see the U.S. Second Infantry Division *stay* in South Korea. Should Washington withdraw it, and South Korean anxiety trigger a new arms race between Seoul and Pyongyang, as it did during the Carter administration, stability would not be served. Certainly both North and South Korea should be talking about ways of reducing arms on the peninsula, and the United States and the Soviets—as the suppliers of the most advanced weapons on the peninsula—need to be associated with these discussions.

Toward Vietnam. The Vietnamese find themselves in a more serious situation than North Korea. Hanoi, by its own choice, is involved in a long-protracted tunnel of conflict in Kampuchea without much light at the end. The end result of the Vietnamese Communists' fifty-year dream to place all of Indochina under their control, Hanoi has been attempting to pacify Kampuchea since December 1978. Having to rebuild the Phnom Penh government from the ground up, Hanoi has found its client policy in Kampuchea to be expensive and tiring. Equally serious has been the growth of Hanoi's dependency on the Soviet Union. In 1984 Hanoi announced its intention, under certain circumstances, to withdraw its forces from Kampuchea by 1990.

Ultimately, however, it may be Vietnam's abject economic situation that pulls Hanoi's troops out of Kampuchea. For example, in late 1985 the Politburo made unusually candid admissions about the extent of Vietnam's economic deterioration since Saigon fell in 1975. "Bureaucratism," wrote Politburo member To Huu in the Communist Party daily *Nhan Dan*, "has driven production and trade installations into passivity, dependence and dullness has not forced them to pay attention to productivity and effectiveness."[34] To boost production and trade, Huu added, dynamic enterprises "must operate secretly and disregard regulations." Huu also confirmed that Vietnam's "per capita national income declined . . . by 10% in 1975 and has continued to decrease annually by 2%–3%."[35] Estimates put Vietnam's current annual per capita income at between $150 to $180. In conjunction with these admissions of failure came Hanoi's announcement that it would soon allow wholly foreign-owned enterprises to set up in Vietnam. With an incentives package due to be implemented in 1986, Hanoi was seeking to build up its extremely low hard currency reserves through increased Western trade, capital flows, and foreign investment. Thus Vietnam is moving to try to follow China's and even North Korea's lead in attracting foreign capital and engaging with non-Communist economies. Then, in late 1986, came startling changes in Hanoi's leadership: Three senior leaders were retired and Nguyen Van Linh, a reputed economic reformer, was chosen Communist Party Secretary.

Accordingly, Vietnam's dire economic situation, its fatigue with the conflict in Kampuchea, and its leadership changes, may open wedges for ASEAN, the United States, and China to affect a moderation in Hanoi's policies. The United States has appropriately associated itself with earlier ASEAN proposals for force deescalation in Kampuchea and moves toward peace and representational government in Phnom Penh. Now, with Vietnam's economy on the ropes, continuing Chinese military pressure on Vietnam's border, China's reminders to the USSR about the need to get the Vietnamese out of Kampuchea, and no victory in sight in Kampuchea, the incentives have mounted for an invigorated allied policy toward Vietnam: take Hanoi up on its withdrawal proposal in return for power sharing in Phnom Penh and economic and political benefits from ASEAN. Hanoi's desire for a normalization with the United States also gives Washington some leverage: no recognition until serious progress is made on the questions of unaccounted-for U.S. forces missing in action and Hanoi's Kampuchean occupation.

The modalities of a cease-fire, withdrawal, and neutralization can be left to ASEAN diplomats working in concert with Hanoi and a Sihanouk-led Cambodian coalition. Indonesia or Malaysia could well become instrumental in such a negotiation: Jakarta, in particular, has good relations with Hanoi, shares Vietnam's concern about China, and acted in a peacekeeping role in South Vietnam during the cease-fire and control period following the 1973 Paris Accords.

Toward Cambodia. Regarding Cambodia, a fundamental weak spot in allied initiatives continues to be the dissension within Prince Sihanouk's tripartite coalition, the Coalition Government for Democratic Kampuchea (CGDK). Bracing for renewed Communist dry season offensives against their resistance forces and base camps, in 1986 Sihanouk and the coalition's two principal deputies, Son Sann and Khieu Samphan, also were having to deal with outbreaks of fighting among the Khmer Rouge faction, and Son Sann's Khmer People's National Liberation Front (KPNLF) forces. Coalition cohesion has been further complicated by renewed leadership struggles within the KPNLF.[36]

Inside Kampuchea, further command and control problems

complicate the resistance's confrontation with the Vietnamese, although most of the guns still belong to the Khmer Rouge, whose total armed forces inside and outside Cambodia may number 35,000 to 40,000, while Son Sann's forces are about 15,000 and Sihanouk's may be 10,000 to 11,500.[37] In December 1985, all three leaders, Sihanouk, Son Sann, and Khieu Samphan, made an official visit to Beijing, where China's leader Deng Xiaoping sought to help them resolve differences. Deng is reported to have told them that China would support their anti-Vietnamese struggle "if it takes a hundred years to succeed."[38] These mediation efforts by members of the U.S.-led coalition need to continue so as to reduce dissonance within Cambodia's resistance forces.

Special Issues

The Philippines. The victory of Corazon Aquino and the democratic transition in the Philippines is proving to be one of the most critical changes affecting the emerging allied security system in Asia and the Pacific. Can the transition be effected in such a way that the forces of democracy, led by Mrs. Aquino and Mr. Laurel, emerge in firm control and better able to deal with the continuing Communist threat to the Philippines? In February 1987, Mrs. Aquino celebrated a year in office, a major referendum victory on her new constitution, and continual strong U.S. backing. But the long-term prospect for real recovery by the Philippines from its political and economic spiral in the last years of Marcos, and the prospects for continued U.S. military operations out of Philippine bases on behalf of Southeast Asia's stability and security, were not clear.

The South China Sea. Offshore Southeast Asia has become the scene of a miniature naval arms race which has the potential to disrupt ASEAN's emerging security response to Communist threats both within and adjacent to the region. The tension involves territorial claims, offshore oil and gas deposits, and other political and judicial problems. Countries with claims in Southeast Asian waters are China, Vietnam, the Philippines, Malaysia, and Taiwan. China, taking into the account the USSR's military deployments out of Vietnam's South China Sea bases, has increased the forces assigned to its South Sea Fleet as well as support facilities along the South China coast, on Hainan Island, and in the Paracels. New Chinese naval/air exercises have occurred in the Tonkin Gulf, the South China Sea, and the Philippine Sea. China claims some 160 islands and reefs in the South China Sea—rocks also claimed by Vietnam, Taiwan, and some ASEAN countries.[39] Clearly to the extent that ASEAN countries and China can mute their own competition regarding these conflicts, they along with the United States stand a better chance of presenting a united front to the USSR.

The Sea of Japan and the Northern Territories. Soviet Communist Party Secretary-General Gorbachev proposed new initiatives in Northeast Asia during his July 1986 Vladivostok speech. But Soviet motives also continue to seek to weaken the emerging allied security coalition in Asia-Pacific and in Northeast Asia in particular. A likely continuing object will be Japan and its relationship with the United States. Shevardnadze's mid-January 1986 trip to Japan, the first by a Soviet foreign minister in ten years, indicated Moscow was trying to drive wedges between Tokyo and Washington on the "Star Wars" issue, although the Japanese proved to be skeptical about Soviet motives. There is potential for a Soviet diplomatic offensive toward Japan. The Soviets seem upbeat about their prospects for improving relations with the Japanese.[40] This also provides some leverage to the Japanese who could, indeed, seek to engage Moscow in arms control negotiations regarding the Sea of Japan and the southern Kurile Islands. Perhaps Prime Minister Nakasone could emulate Deng Xiaoping's approach by offering negotiations with Gorbachev, *provided* he shows real flexibility on the military and territorial issues which separate the two countries.

In the remainder of this book some of the aforementioned conflict patterns and strategies for coping with threats to regional security and stability will receive further analysis. The seven individual chapters will focus on the substantive issues and emerging trends in each of seven East Asian conflict zones: the Sino-Soviet conflict, the Japanese-Soviet northern territorial dispute, stability and instability in the Sea of Japan, the Korean peninsula zone of conflict, the South China Sea, Thai-Vietnamese rivalry in the Indochina conflict, and Philippine communism and its threat to Philippine democracy and stability.

The book closes with a concluding chapter which explores some of the policy measures necessary for deescalating intraregional conflict and for eventual resolutions of conflict issues in the East Asia–Western Pacific region. The discussion will seek ways of steering the region's increasingly complex conflict patterns into specific zone-by-zone negotiations, designed to reduce forces and tensions so as to promote peace and stability.

Notes

1. See, for example, Roy Hofheinz, Jr., and Kent E. Calder, *The Eastasia Edge* (New York: Basic Books, 1982).
2. Richard H. Solomon and Masataka Kosaka, eds., *The Soviet Far East Military Buildup: Nuclear Dilemmas and Asian Security* (Dover, Mass.: Auburn House Publishing Co., 1986).
3. Michael H. Armacost, "The Asia-Pacific Region: A Forward Look," U.S. Department of State, *Current Policy*, No. 653, 29 January 1985. See also Gaston J. Sigur, Jr., "U.S. and East Asia-Pacific Relations: The Challenges Ahead," U.S. Department of State, *Current Policy*, No. 859, 14 July 1986.
4. Donald S. Zagoria, "The Soviet Peace Offensive in the Pacific Basin," a presentation before the National Defense University Pacific Basin Security Symposium, Honolulu, Hawaii, February 27, 1987. The Soviets' major partners being China, Japan, North Korea, Vietnam, and Mongolia. Soviet trade with the rest of the area, particularly Southeast Asia and the South Pacific, is minimal, New Zealand being the sole exception. See, for example, Paul Dibb and T. B. Millar, eds., "The Interests of the Soviet Union in the Region: Implication for Regional Security," *International Security in the Southeast Asian and Southwest Pacific Region* (St. Lucia, Queensland: Queensland University Press, 1983), pp. 19–46.
5. *Asia Week*, 12 January 1986, p. 26.
6. George P. Shultz, "Challenges Facing the U.S. and ASEAN," U.S. Department of State, *Current Policy*, No. 597, 13 July 1984. See also George P. Shultz, "Pacific Tides Are Rising," address to the World Affairs Council of Northern California in San Francisco on 5 March 1983; George P. Shultz, "The U.S. and ASEAN: Partners for Peace and Development," U.S. Department of State, *Current Policy*, No. 722, 12 July 1985.
7. T. B. Millar, "Introduction: Asia in the Global Balance," in Donald Hugh McMillen, ed., *Asian Perspectives on International Security* (London: Macmillan, 1964), p.3.

8. *Ibid.*, p. 7.
9. *Ibid.*, pp. 5–6.
10. For details, see Young Whan Kihl and Lawrence E. Grinter, "New Security Realities in the Asian-Pacific," in Young Whan Kihl and Lawrence E. Grinter, eds., *Asian-Pacific Security: Emerging Challenges and Responses* (Boulder, Colo.: Lynne Rienner Publishers, 1986), pp. 4–6. See also Alan D. Romberg, "New Stirrings in Asia," *Foreign Affairs*, Vol. 64, No. 3 (1986), pp. 517.
11. Solomon and Kosaka, eds., *The Soviet Far East Military Buildup*, p. 272; Harry Gelman, "The Soviet Far East Military Buildup: Motives and Prospects," *ibid.*, pp. 41–42.
12. Solomon and Kosaka, eds., *ibid.*, p. 272; Kihl and Grinter, eds., *Asian-Pacific Security*, pp. 4, 54.
13. For excerpts from Gorbachev's speech, see *New York Times*, 29 July 1986. The full text of Gorbachev's speech appears in *The Current Digest of the Soviet Press*, Vol. 38, No. 30 (27 August 1986), pp. 1–8, 32.
14. *New York Times*, 29 July 1986.
15. The notion of a Soviet "comeback" in Asia is from Donald Zagoria. See Donald Zagoria, ed., *Soviet Policy in East Asia* (New Haven, Conn.: Yale University Press, 1982). See also his "The USSR and Asia in 1985: The First Year of Gorbachev," *Asian Survey*, Vol. 26, No. 1 (January 1986), pp. 15–29, and Romberg, "New Stirrings in Asia," *op. cit.*, pp. 525–526.
16. *Far Eastern Economic Review*, 14 August 1986, pp. 36–37.
17. *Christian Science Monitor*, 31 July 1986; 11 August 1986; 19 August 1986, p. 9; 9 September 1986, p. 9.
18. *Ibid.*, 9 September 1986, p. 9.
19. This material draws on Lawrence E. Grinter, "The United States in East Asia: Coping with the Soviet Build Up and Alliance Dilemmas," in Kihl and Grinter, eds., *Asian-Pacific Security*. See also Secretary of Defense Caspar W. Weinberger, "The Five Pillars of Our Defense Policy in East Asia and the Pacific," *Asia-Pacific Defense Forum*, Vol. 9, No. 3 (Winter 1984–85), pp. 2–8.
20. With about 160,000 U.S. combat forces on shore and afloat in the region, only Brunei, Singapore, Laos, Kampuchea, Malaysia, New Zealand, and Australia have fewer forces in East Asia. See *The Military Balance, 1984–1985*, as reprinted in *Pacific Defense Reporter*, 1985 Annual Reference Edition (December 1984–January 1985), pp. 137–147. The 160,000 U.S. combat, or combat support, troops in the Asian-Pacific area are composed mainly of Army divisions in Korea and Hawaii, a Marine division and brigade in Okinawa and Hawaii, Seventh Fleet assets, and U.S. Air Force strategic and tactical fighter

squadrons. See Caspar W. Weinberger, *Annual Report to the Congress, Fiscal Year 1986* (Washington, D.C.: GPO, 4 February 1985), pp. 237–240.
21. See, for example, Zhang Jia-Lin, "The New Romanticism in the Reagan Administration's Asian Policy: Illusions and Reality," *Asian Survey*, Vol. 24, 10 November (October 1984), pp. 1008–1009.
22. In particular, see Alexander M. Haig, Jr., *Caveat: Realism, Reagan, and Foreign Policy* (New York: Macmillan, 1984), pp. 194–217.
23. It is also obvious, however, that neither side will be taken for granted as, for example, the thorny issues of ship visits, Taiwan, and abortion policy indicate. See Richard Baum, "China in 1985: The Greening of the Revolution," *Asian Survey*, Vol. 26, No. 1 (January 1986), pp. 48–51, and Romberg, "New Stirrings in Asia," *op. cit.*, pp. 526–529.
24. Prime Minister Nakasone, while working to eliminate the 1 percent of GNP defense expenditure ceiling, continues to be blocked by LDP senior leaders and the inertia of public opinion. Susan J. Pharr, "Japan in 1985: The Nakasone Era Peaks," *Asian Survey*, Vol. 26, No. 1 (January, 1986), p. 61.
25. Long interviewed by Neil Ulman and Urban C. Lehrer in "Tokyo's Buildup," *Wall Street Journal*, 22 November 1982, p. 1.
26. Frank Langdon, "Japan and North America," in Robert S. Ozaki and Walter Arnold, eds., *Japan's Foreign Relations* (Boulder, Colo.: Westview Press, 1985), p. 29.
27. Japanese Defense Agency (JDA), *Defense of Japan, 1984* (Tokyo: *Japan Times*, 1984), pp. 31–37, and JDA, *Defense of Japan, 1985* (Tokyo: *Japan Times*, 1985), pp. 24–32.
28. Tai Sung An, *North Korea in Transition: From Dictatorship to Dynasty* (London: Greenwood, 1983), p. 81, and C. I. Eugene Kim, "Civil-Military Relations in the Two Koreas," *Armed Forces and Society*, Vol. 11, No. 1 (Fall 1984), p. 12. See also *The Military Balance, 1985–1986* (London: International Institute for Strategic Studies, August 1985), pp. 126–127.
29. See C. I. Eugene Kim, "South Korea in 1985: An Eventual Year Amidst Uncertainty," *Asian Survey*, Vol. 26, No. 1 (January 1986), pp. 67–72.
30. See Donald Weatherbee's chapter in Kihl and Grinter, eds., *Asian-Pacific Security*.
31. See, for example, Mikhail S. Berstam, "Soviet Oil Woes: Detente on U.S. Terms?", *Wall Street Journal*, 10 January 1985, p. 16.
32. For background, see Young Whan Kihl, *Politics and Policies in Divided Korea: Regimes in Contest* (Boulder, Colo.: Westview Press, 1984).

33. Young Whan Kihl, "The 'Hermit Kingdom' Turns Outward," *Asian Survey*, Vol. 25, No. 1 (January 1985), pp. 65–79, and Young Whan Kihl, "North Korea's New Pragmatism," *Current History*, Vol. 5, No. 510 (April 1986), pp. 164–167, 198.
34. To Huu writing in *Nahn Dan* as cited in *Asia Week*, 29 November 1985, p. 22.
35. *Ibid*.
36. William J. Duiker, "Vietnam in 1985: Searching for Solution," *Asian Survey*, Vol. 26, No. 1 (January 1986), pp. 103–104.
37. *Asia Week*, 19 January 1986, pp. 27, 33. See also Michael Eiland, "Cambodia in 1985: From Stalemate to Ambiguity," *Asian Survey*, Vol. 26, No. 1 (January 1986), pp. 120–121.
38. *Asia Week*, 19 January 1986, p. 26.
39. Bradley Hahn, "South-East Asia's Miniature Naval Arms Race," *Pacific Defence Report*, September 1985, p. 22.
40. Zagoria, "The USSR and Asia in 1985: The First Year of Gorbachev," *Asian Survey, op. cit.*, p. 18.

2. Sino-Soviet Relations in the Late 1980s: An End to Estrangement?

Steven I. Levine

ON JULY 28, 1986, in his first major pronouncement on Soviet Asian policy since coming to office sixteen months earlier, Soviet Communist Party General-Secretary Mikhail Gorbachev made a broad-ranging series of proposals aimed at improving Soviet relations with its Asian neighbors. Reaffirming the USSR's status as an Asian-Pacific nation, Gorbachev stated his support for China's current modernization program and called upon Chinese leaders to build on the advances in Soviet-Chinese relations achieved over the past few years.[1] In a September 2 interview with CBS Television, China's preeminent leader Deng Xiaoping responded to Gorbachev's initiative by saying that he would be willing to visit Moscow provided the USSR ceased helping Vietnam in its occupation of Cambodia.[2] What do these top-level declarations indicate about the state of Sino-Soviet relations?

As early as the summer of 1985, in the wake of Chinese Vice-Premier Yao Yilin's successful visit to Moscow, which resulted in the signing of several important economic cooperation agreements, the influential Chinese journal *Liaowang (Outlook)* asserted: "The years of estrangement in Sino-Soviet relations are now over."[3] Considerable evidence is available to sustain such an assessment. Exchanges of visits by high-ranking leaders, burgeoning trade, the intensification of cultural and educational exchanges, and a marked improvement in the general atmosphere have characterized Sino-Soviet relations over the past couple of years. The prospects for further improvement appear good.

The turnaround in Sino-Soviet relations has occurred incrementally, without any of the dramatic moments that characterized the improvement of Sino-American relations fifteen years earlier, but the significance of this change for international relations both regionally and globally is undeniable. A backward glance both at the origins of the Sino-Soviet conflict and the initial stages of the rapprochement may help to put the current stage of the relationship in better perspective.

I. Origins of the Sino-Soviet Conflict

Twenty-five years after the collapse of the Sino-Soviet alliance, the origins of the conflict between the two major Communist powers remain insufficiently understood. Elements of the complex causality include clashes of national interest, divergent foreign policy objectives, ideological disagreements, and personal antagonism between Chinese leaders (especially Mao Zedong) and Soviet leaders (especially Nikita Khrushchev). Underlying all these discrete causes, as I have argued elsewhere, was a structure of unfulfilled and unrealistic expectations that each side entertained with respect to the other at the beginning of their alliance.[4] Chief among these was the Chinese view that the USSR really could function as a selfless and generous elder brother willing to nurture the Chinese younger brother and share power in the international Communist movement. Moscow's expectation that Beijing would be content to remain a loyal and subordinate member of the Soviet camp while pursuing a Soviet-style path of development was equally flawed.

Rooted as it was in a concept of politics (proletarian internationalism) that denied the possibility of legitimate conflicts among socialist nations, the Sino-Soviet alliance failed to develop the conflict resolution mechanisms critical to successful long-term alliance relationships. Sharing a zero-sum Leninist-style of conflict resolution, Soviet and Chinese leaders quickly escalated their specific points of disagreement, and staked out mutually opposed positions from which they seemed unable to retreat short of surrender, especially given the investment of personal political prestige by leaders on both sides. The small-scale border war of spring 1969 and continuing volleys of polemics appeared to set the Sino-Soviet conflict in concrete. It is important to note, however, that by the autumn of 1969, the risk of an all-out Sino-Soviet war had already receded and the two sides had commenced a series of talks which, without achieving specific results, succeeded in defusing tension somewhat, much like the Sino-American ambassadorial talks in Geneva and Warsaw in the 1950s and 1960s.

II. The Road to Sino-Soviet Detente

The logic of Sino-Soviet tension reduction was evident to many observers well before the actual process commenced in earnest. Throughout the 1970s, Chinese leaders rejected or ignored repeated Soviet efforts to reach agreement on specific tension reduction agreements. Chinese foreign policy was directed toward encouraging the formation of a global anti-Soviet coalition at the same time that Beijing moved to establish and consolidate its relations with the West. Shortly after the establishment of diplomatic relations with the United States in early 1979, China signaled its desire to open up a new channel of discussion with the USSR on the issues outstanding between them. (At the same time, the thirty-year Sino-Soviet Treaty of Friendship and Alliance, which had long been a dead letter, was allowed to lapse.) Talks between Moscow and Beijing, which began in the fall of 1979, were suspended but not broken off by the Chinese side in response to the Kremlin's invasion of Afghanistan in December.

By 1981–82, Beijing was in a mood more receptive to the idea of Sino-Soviet tension reduction. Sino-American relations were in a state of temporary decline as the issue of U.S. arms sales to Taiwan clouded the horizon, and the pace of U.S.-PRC strategic cooperation visibly slowed. Distancing itself from the United States, Beijing resumed its condemnation of the "hegemonistic behavior" of both superpowers and proclaimed an independent foreign policy more in keeping with the historic spirit of the Chinese nation than the de facto tilt toward Washington of the preceding several years.

It was in this atmosphere that Soviet leader Leonid Brezhnev, in the last year of his life, undertook an initiative to break the Sino-Soviet stalemate. In a March 1982 speech in the Central

Asian city of Tashkent, Brezhnev reiterated Soviet interest in confidence-building measures along the Sino-Soviet border, and he called for the improvement of relations on the basis of "mutual respect for each other's interests, non-interference in each other's affairs, and mutual benefit."[5] He reiterated this position just a few weeks before his death in a September 1982 speech in Baku. This time Beijing responded favorably to the Soviet suggestion that the suspended talks be resumed, and in October 1982 the first of an ongoing series of vice-ministerial consultations commenced. These talks continue to this day, alternating between Moscow and Beijing about every six months.

Before examining the most recent period in Soviet-Chinese relations, it may be pertinent to raise the question of what "normalization" actually means in the context of Sino-Soviet relations. In the thirty-eight years since the establishment of the People's Republic of China in 1949, Sino-Soviet relations have run the gamut from the "eternal friendship" of the 1950s through the "permanent enmity" of the late 1960s and 1970s. The wildly inconsistent history of Sino-Soviet relations thus provides little guidance in defining normalization or normality.

Nor is the parallel history of Sino-American relations of much help. In the Washington-Beijing relationship, "normalization" was defined by both sides as the establishment of full diplomatic relations, a task accomplished on January 1, 1979. In the Sino-Soviet case, even during the periods of greatest tension and hostility, the framework of diplomatic relations established in 1949 was not splintered, trade continued at very low levels, and Moscow never challenged Beijing's claim to sovereignty over Taiwan. During the Sino-Soviet split, the most decisive break occurred in the realm of party-to-party relations, and no significant steps have been taken as of this writing to repair the break in that area.

III. Trade and Economic Relations

Trade and economic relations have thus far been the major instruments for overcoming Sino-Soviet estrangement, although the use of economic relations derived from larger political considerations in both capitals. In the first stages of the post-Mao economic revitalization program, China multiplied its links with the developed capitalist world very rapidly, while making only modest progress in its trade with the Soviet Union and the countries of Eastern Europe. In 1982, Beijing decided to include the Soviet Union and Eastern Europe in its Open Door economic diplomacy. In addition to the economic benefits of this decision, Chinese leaders must have calculated that the expansion of trade and economic ties with the Soviet Union would enhance the credibility of Beijing's newly proclaimed independent foreign policy.

For its part, Moscow had watched growing Western involvement in China's economy with considerable unease. Soviet leaders had long called for an increase in economic relations with China to help reduce tension. Thus Moscow responded with alacrity to China's 1982 decision, seeing trade as a way to improve the overall Sino-Soviet relationship.

The new Sino-Soviet economic relationship was forged in meetings at the vice prime ministerial level, beginning in December 1984. That month, Soviet Vice Prime Minister Ivan V. Arkhipov, an economic affairs specialist who had headed the Soviet economic assistance program to China in the 1950s, returned to Beijing; he was the highest ranking official Soviet visitor in fifteen years. Greeted warmly as an old friend of China's, Arkhipov held substantive talks with his Chinese counterpart, Chinese Vice Prime Minister Yao Yilin, and was received by Peng Zhen, chairman of the Standing Committee of the National People's Congress, and by other prominent leaders of the older generation.

At the end of the visit, several agreements were signed that provide for: (1) economic-technical cooperation in modernizing industrial enterprises developed in the 1950s with Soviet aid; (2) scientific and technical exchanges; (3) the establishment of a Sino-Soviet Economic, Trade, Scientific, and Technical Cooperation Commission; and (4) a long-term trade agreement (1986–90) envisioning a rapid increase in trade.[6]

In July 1985, Yao Yilin reciprocated Arkhipov's visit by traveling to Moscow, where he signed agreements concerning trade and payments for the 1986–90 period. The volume of trade for this period was set at $14 billion. Another agreement specified Soviet assistance in the construction of seven new economic projects and the reconstruction of seventeen older facilities in fields like machine building, metallurgy, coal and chemical production and transportation.

Sino-Soviet trade has expanded rapidly over the past several years. As recently as five years ago, the total value of the two-way trade was only about $160 million. By 1984, trade had increased to $1.2 billion, and in 1985 it took a further leap to $1.9 billion. However, this was still only slightly over 3 percent of China's total foreign trade, and only 30 percent of the value of China's trade with the United States.[7] When the long-term trade agreement expires in 1990, Soviet-Chinese trade is projected to have grown to $6 billion, somewhat less than the U.S.-China trade figure for 1985. However, it seems probable that the actual figures for Sino-Soviet trade will outstrip these projections. Chinese exports to the Soviet Union are mostly agricultural and light industrial products, including foodstuffs, handicrafts, textiles, and minerals, while the Soviet Union ships China machinery, steel, electrical power equipment, fertilizer, transportation equipment, and other heavy industry products. As the Soviet nonestablishment scholar Roy Medvedev has pointed out, the underdevelopment of the Soviet consumer industry makes the Soviet Union a natural market for Chinese light industrial products, without having to face the protectionist barriers of Western markets.[8]

A notable feature of Sino-Soviet economic relations since 1983 has been the renewal of the once vigorous trade linking contiguous regions in China and the Soviet Union on a barter basis. From Xinjiang in the far northwest of China through Heilongjiang in the Northeast, new border trading posts have been opened and the volume of trade has expanded rapidly.

In March 1986, Vice Prime Minister Arkhipov returned to Beijing to chair the first session of the Sino-Soviet Economic, Trade, Scientific, and Technical Cooperation Commission and to sign another protocol on the exchange of engineers and technicians. *Pravda*'s report on this visit noted that Chinese Prime Minister Zhao Ziyang, who met with Arkhipov, expressed satisfaction with progress in trade, economic, technical, and scientific cooperation between the Soviet Union and China. However, the Soviet party newspaper did not publish Zhao's complaint that no substantial progress had been achieved in political relations.[9]

The implementation of the agreements signed by Vice Prime Ministers Arkhipov and Yao required the multiplication of Sino-Soviet contacts at the working level and the reanimation of links that had lain dormant for twenty years or more. In May 1986, for example, a delegation of the Soviet Ocean Shipping Company visited Shanghai and reached an agreement to establish an office in that city to handle the growing volume of trade. A counterpart Chinese office is to be set up in Odessa.

After a twenty-year gap, river transport between the Soviet Union and China resumed along the Heilongjiang (Amur) and Songjiang rivers linking northeast China and eastern Siberia. China's first major trade exhibition in the Soviet Union since 1953 took place in July–August 1986. The reawakening of Sino-Soviet relations brings to mind Washington Irving's story of Rip Van Winkle. Like the hapless Dutchman, who returned to his village after a twenty-year sleep to find his world transformed, the Soviet Union and China are resuming their intercourse after a long hiatus. But the world of the 1950s is gone forever.

IV. Cultural Relations and Contacts

During the process of Sino-American normalization, the establishment of high-level contacts was followed by the "thickening" of the Sino-American relationship through the multiplication of economic, cultural, educational, tourist, and other links. A similar process is now under way in Sino-Soviet relations. But unlike China's relations with the United States, where a multitude of private American organizations, businesses, and individuals have established links with the Chinese, in the Sino-Soviet arena all the strands of the relationship have an official or quasi-official character.

The growth of educational exchanges provides one barometer of Sino-Soviet cultural relations. Starting with only ten students from each side in 1983–84, there were seventy in the following year, and currently there are over two hundred exchange students. To put this in perspective, however, one should observe that this figure is equal to only a little over 1 percent of the number of Chinese students from the PRC in the United States at the same time. After a lapse of twenty years, Soviet and Chinese artists, musicians, dancers, athletes, filmmakers, and others are again performing in one another's country under the terms of cultural cooperation agreements signed by the two governments.

The reanimation of Sino-Soviet cultural, educational, scientific, and technical exchanges has occurred without any obvious problems. Although Soviet culture does not inspire the same enthusiasm in Chinese urban youth that Western popular culture does, it does not carry the risk of "spiritual pollution" that Chinese cultural conservatives see lurking in "decadent capitalism." Nor do Soviet and Chinese officials fear politically inspired defections like the celebrated defection of Chinese tennis star Hu Na, which caused a minor crisis in Sino-American relations in 1983.

In sum, the prospects for the broadening and deepening of Sino-Soviet cultural relations are good within the limits established by officials on both sides. The renewal of Sino-Soviet cultural relations and the growing if still modest contacts between Chinese and Soviet citizens in various walks of life give the lie to one of the less attractive myths engendered by the Sino-Soviet conflict, namely, that deep-seated historical and cultural antagonisms verging on race hatred lay at the root of the Sino-Soviet conflict.

V. Political and Military Relations

An analysis of the overtly political dimensions of the Sino-Soviet relationship shows signs of improvement alternating with signs of continuing conflict and clashes of interest. Factoring in the military dimension adds further to the somber side of the relationship. The election of Mikhail Gorbachev as General Secretary of the Communist Party of the Soviet Union (CPSU) in March 1985, after the death of Konstantin Chernenko, allowed both sides to reiterate their commitments to improving relations. In his maiden speech as General Secretary to a special Central Committee plenum on March 11, 1985, Gorbachev said, "We would like to see a serious improvement in relations with the People's Republic of China, and believe that, given reciprocity, this is quite possible."[10]

China's National People's Congress Chairman Peng Zhen praised Chernenko's dedication to the improvement of Sino-Soviet relations and, echoing Gorbachev's call, said, "We too cherish the same hope. The Chinese Government will do its best to constantly develop Sino-Soviet relations in various fields."[11]

Vice Prime Minister Li Peng, a top leader of the younger generation (and a man educated in the Soviet Union), headed the Chinese delegation to Chernenko's funeral. In Moscow, he met with Gorbachev and reaffirmed China's commitment to improved relations with the Soviet Union. (A lower-ranking Chinese official, then Foreign Minister Huang Hua, had represented China at Soviet President Leonid Brezhnev's funeral in November 1982.) General Secretary Gorbachev's assessment of Sino-Soviet relations in his report to the Twenty-Seventh Soviet Party Congress on February 25, 2986, was upbeat.[12]

To Chinese observers, the program of economic revitalization and political renewal that Gorbachev promised upon assuming office, which involved the wholesale removal of elderly holdovers from the Brezhnev era, may have appeared to be the Soviet equivalent of Chinese leader Deng Xiaoping's reforms. In any case, Chinese press commentary on Soviet domestic affairs in the months after Gorbachev took power tended to be objective and noncritical in tone. One stimulus to China's greater interest in improving Sino-Soviet relations, then, may have been the realization that Gorbachev would be a more effective and dynamic leader. Beijing may have expected some new initiative from Gorbachev to break the stalemate over the so-called Three Obstacles. Such an initiative finally came in the July 28, 1986, speech which will be analyzed below. But in the first year plus of Gorbachev's tenure, domestic reform and relations with the United States took priority over Sino-Soviet relations. From a Soviet perspective, existing trends in the relationship were encouraging and there was no need to offer the Chinese any substantial concessions.

Even in the strictly political realm, Sino-Soviet relations advanced markedly in 1985 and 1986. In March 1985, a delegation of Chinese representatives to the National People's Congress led by Zheng Chengxian traveled to Moscow in the first such visit in more than twenty years. In October, a reciprocal Soviet parliamentary visit to China took place, led by Lev N. Tolkunov of the Supreme Soviet of the USSR. The Soviet group met with President Li Xiannian, Peng Zhen, and other important Chinese officials, amid the by now familiar expressions of determination to work for even better relations. In September 1985, Soviet Foreign Minister Eduard Shevardnadze and his Chinese counterpart Wu Xueqian met at the United Nations' annual General Assembly session to exchange points of view on international issues in what promised to become an annual encounter. Indeed, the exchange of views took place once more in the fall of 1986.

Thus, at the diplomatic level, the number of high-level contacts between the Soviet Union and China increased substantially in 1985–86. Multiple channels of communication were open to officials in many spheres of activity, and only the very highest level visits were considered out of the ordinary. Setting aside national security concerns for the moment, in the sphere of strictly bilateral relations there were no outstanding conflicts between the two countries that harbored the seeds of crisis, nor were there any such strictly bilateral issues that required the urgent attention of a Gorbachev or a Deng Xiaoping.

If this assertion is correct, can it then be concluded that Sino-Soviet relations have been normalized? It must be remembered that the Chinese vociferously reject such an idea. Against the melody of Sino-Soviet amelioration, Chinese officials sound the bass refrain that the relationship cannot be normalized so long as Soviet leaders refuse to budge on the Three Obstacles.

A specific case (as of the end of 1986) is the Chinese refusal to reestablish party-to-party links between the CCP and the CPSU. In October 1985, Wu Xingtang, a spokesman for the international liaison department of the Chinese Communist Party, said it was not yet time to consider restoring relations between the Soviet and Chinese Communist parties in view of the continuing security threat to China posed by the Three Obstacles. Just six months later, a Chinese party spokesman repeated the same point. Repeating a long-established practice, the CCP refrained from sending a delegation to the Twenty-Seventh Congress of the CPSU held in Moscow in February 1986. Twenty-five years have passed since China's Prime Minister Zhou Enlai demonstratively stalked out of the Soviet Twenty-Second Party Congress to protest Prime Minister Nikita Khrushchev's verbal onslaught against Albania and the ghost of Joseph Stalin. In the area of party-to-party relations, Rip Van Winkle is still sleeping.

Why have the Chinese thus far been unwilling to resume formal party relations with the CPSU? After all, CCP Politburo member Li Peng has already met twice with Gorbachev; other officials involved in state-to-state relations are, of course, highly placed in their respective Communist parties. Furthermore, in recent years Chinese leaders have stressed the desirability of establishing links with so-called progressive parties of widely varying political persuasions. Why not the Soviet Communist Party?

Chinese leaders adduce the famous Three Obstacles as standing in the way of the resumption of Soviet and Chinese Communist Party ties. These Three Obstacles are the concentration of Soviet forces along the Chinese border (including Soviet forces in Mongolia); Soviet intervention in Afghanistan; and Soviet support for Vietnam's occupation of Cambodia. When one reflects upon these Three Obstacles, it is apparent that they have nothing whatsoever to do with party relations. Why, then, should formal party relations be held hostage to significant changes in Soviet foreign and security policy—changes that Chinese leaders may demand but cannot reasonably hope to effect through any action on their own? Are party-to-party relations so important that they are being held in reserve as the last area to be normalized? Or is it perhaps that formal party relations matter so little in practice that the Chinese at a minimum lose nothing and perhaps have something to gain in postponing this final step?

In Marxist-Leninist terms, of course, the party stands above the state structure in the hierarchy of political power, and is the repository of ultimate authority and legitimacy. By abstaining from reestablishing relations with the CPSU, then, the Chinese Communists can indicate their continuing disapproval of Moscow's policies in the areas of foreign and security policies without jeopardizing the concrete interests that are served by the reforging of the links. Moreover, the core value of political autonomy is symbolically protected by China's refusal thus far to reestablish party links.

This becomes clear if one recalls the origins of the Sino-Soviet conflict. A central element in that conflict was China's refusal to accept a subordinate position in the world Communist movement and its determination to contest the leadership of that movement (which had remained under Moscow's domination even after de-

Stalinization).[13] One of the fixed points in the recent shifts in Chinese policies is the idea that relations among Communist and other "progressive" parties should be guided by "the principles of independence, equality, mutual respect and non-interference in each other's internal affairs."[14] Abundant evidence sustains the Chinese assertion that the Soviet Communist Party fails to respect these principles in its dealings with smaller and weaker parties in the socialist world. Until the Soviet Union (directed by the CPSU) ceases to assert hegemony in its dealings with other Communist parties and socialist states, the CCP, according to the logic of principle, cannot enter into relations with the CPSU.

Let us return, then, to the Three Ostacles that (the Chinese say) impede the normalization of Sino-Soviet relations. Between the fall of 1982 and the fall of 1986 nine rounds of Sino-Soviet consultations were held at the vice foreign ministerial level. These meetings, alternating between Moscow and Beijing about every six months, have become institutionalized as a channel of Chinese and Soviet contact. After these meetings, Chinese Foreign Ministry spokesmen routinely repeat China's complaint that the Soviet Union is unwilling to take concrete steps to remove the Three Obstacles from the road of Sino-Soviet normalization. In his foreign policy report to the fourteenth session of the sixth National People's Congress on January 16, 1986, Foreign Minister Wu Xueqian said that despite some improvements in Sino-Soviet relations in 1985, "no fundamental improvement has ever been in sight in the political relations between the two countries." Wu asserted that Soviet officials should confront the Three Obstacles rather than avoiding them, and he suggested that the first priority should be "for the Kremlin to stop supporting Vietnam in its aggression against Kampuchea."[15]

It is quite clear that the continuing clash of Soviet and Chinese international political and geostrategic interests lies at the base of the unresolved Three Obstacles. Despite the improvements in Sino-Soviet relations, PRC leaders still view the USSR as the primary threat to China's national security, even if they no longer fear an imminent military attack. Chinese leaders are well aware of the massive Soviet land, air, and naval deployments in the Far East, including an estimated 52 divisions of troops, approximately

160 mobile SS-20 IRBMs, and a Pacific Fleet numbering some 88 submarines and some 300 other surface ships.[16] The Kremlin's access to bases in Mongolia and in Vietnam threatens Chinese security from both the northern and southern peripheries. Against these forces, the bulk of the People's Liberation Army, deficient in modern military technology, is deployed in the military regions adjacent to the USSR. Despite the beginnings of force modernization (aided by modest transfers of Western military technology), it will be a long time indeed, if ever, before the Chinese can aspire to anything approaching an effective balance of forces vis-à-vis the USSR.

Should one understand the meaning of the Three Obstacles, then, in terms of the clash between Soviet and Chinese security interests? It is certainly true that these interests clash in Afghanistan and Cambodia. The Soviet goal of pacifying Afghanistan is frustrated, in part, by Chinese support of the resistance movement. Peking's goal of loosening Vietnam's grip on Cambodia and promoting Chinese influence in Indochina is frustrated by Moscow's strong support of Hanoi. Soviet and Chinese officials continue to trade bitter charges on these issues. Both conflicts are already long-drawn-out affairs that are unlikely to be settled in the near term, but neither involves the vital interests of the Soviet Union or China. (Although the Soviet-Vietnamese alliance may enhance Moscow's military encirclement of China, it is the issue of Vietnam's involvement in Cambodia rather than the Moscow-Hanoi axis itself that Beijing includes as one of the Three Obstacles.)

Moscow has repeatedly stated that it will not normalize relations with Beijing at the expense of third parties, although the recent Gorbachev initiatives suggest some flexibility on that point. Nevertheless, until solutions to the Cambodian and Afghan issues are arrived at outside of the arena of Sino-Soviet relations, two of the Three Obstacles will remain. The issue of Soviet troop concentrations along the Chinese border is more amenable to face-saving diplomatic solutions; the Soviet Union could reduce the strength of some units and relocate others without substantially reducing its strategic advantage vis-à-vis the PRC. In any case, as already noted, although China continues to upgrade its military capabilities, the Chinese no longer express anxieties about an imminent Soviet military threat. Substantial numbers of Chinese officers and troops have been demobilized and defense industries shifted partially to civilian production.

The key to the meaning of the Three Obstacles, I would suggest, may be found in China's overall foreign policy posture, particularly its stance toward the superpowers. Since 1982, China has proclaimed an independent foreign policy, which eschews strategic relations or alliances with any large power or bloc of powers. Terming both the United States and the Soviet Union "hegemonists," the Chinese criticize aspects of both Soviet and American foreign policy even while developing relations with both of them. In the relationship with the United States, the question of Taiwan has both real and symbolic importance; it is the issue Washington and Beijing have failed to resolve. In Alfred Wilhelm's perfect metaphor, the Taiwan issue is the nuclear control rod that Beijing raises and lowers to control the temperature of Sino-American relations.[17]

In Sino-Soviet relations, the Three Obstacles perform an analogous function. Beijing's insistence that the Soviet Union remove the Three Obstacles is a kind of symbolic assertiveness that enables the Chinese to enhance their sense of autonomy even as they intensify their economic cooperation and cultural exchanges with the USSR. The Three Obstacles (like the Taiwan issue) guard against the Chinese proclivity to fall into the arms of one or another external patron only to recoil later in frustration and anger. They are a back brace for China's independent foreign policy. The Three Obstacles also reassure the United States and others in the West about the limits of Sino-Soviet rapprochement. In this sense, the Three Obstacles are not a barrier but a Chinese screen to shield the amelioration of Sino-Soviet relations. Meanwhile, Americans can continue to develop their relations with the Chinese, secure in the belief that Sino-Soviet relations can progress only until they run aground of the Three Obstacles.

If this interpretation makes any sense, what are the chances for success of General Secretary Gorbachev's initiative of July 1986 toward the Chinese? As noted at the beginning of this chapter, Gorbachev's speech in Vladivostok represented a major Soviet effort to introduce a greater element of Soviet flexibility into the

Sino-Soviet dialogue as a means of breaking the impasse over the Three Obstacles. Gorbachev touched upon two of the Three Obstacles by promising a partial Soviet withdrawal from Afghanistan (generally dismissed in the West as an empty gesture), and by raising the possibility of withdrawing Soviet forces from Mongolia and, in cooperation with the Chinese, from along the Sino-Soviet border itself. He also proposed joint water management and economic cooperation for the Amur (Heilongjiang) River Basin, reiterated an earlier Soviet commitment to treating the main channel of the Amur as the international boundary between the USSR and the PRC, and revived the concept of a rail link between Soviet Kazakhstan and Chinese Xinjiang.[18]

China's initial reaction to Gorbachev's proposal was rather cautious, emphasizing the need for detailed study before making a commitment. This attitude suggests that careful consideration was given in Beijing as to how to exploit the positive elements in Gorbachev's proposal without conceding too much of an international propaganda advantage to the Soviets. In a meeting with the Soviet chargé d'affaires in Beijing, Foreign Minister Wu Xueqian welcomed Gorbachev's speech, but said that the General Secretary had not gone far enough to remove the Three Obstacles. In particular, he drew attention to the fact that the General Secretary had not addressed the issue of Soviet support for Vietnam's occupation of Cambodia.[19]

Finally, it was left to Deng Xiaoping, China's feisty octogenarian leader, to come up with an appropriate riposte to Gorbachev. Speaking to Mike Wallace of the CBS News program "60 Minutes," Deng offered to travel to Moscow for a summit meeting with Gorbachev provided that the Soviets ceased supporting Vietnam's occupation of Cambodia. In effect, Deng indicated that only one of the Three Obstacles—conspicuously the one the Soviet leader had failed to address—was the real barrier to Sino-Soviet normalization. Soviet flexibility was made to appear as inflexibility, and the Chinese retained their principled opposition to full political relations with Moscow.

Nevertheless, events in the autumn of 1986 suggested that Chinese opposition to party-to-party relations with the CPSU might be eroding. The heads of the Communist parties of Poland and the German Democratic Republic—General Jaruzelski and Erich

Honecker, respectively—were welcomed to Beijing, where they met with top CCP officials. At an earlier period, Beijing's flirtation with the maverick Rumanian Communists was part of an effort to drive a wedge between Moscow and its East European allies. Now the welcome extended to the Poles and the East Germans presaged a general relaxation of tension with the Soviet bloc, including Moscow itself.

VI. Implications of Sino-Soviet Amelioration for the United States

What are the implications of the end of Sino-Soviet estrangement for the United States? Are legitimate American interests threatened by the present level of Sino-Soviet relations or are they likely to be endangered by further progress toward Sino-Soviet rapprochement? Some tentative thoughts with regard to these questions are in order.

For a long time, American officials and observers tended to see politics in the so-called strategic triangle of Sino-Soviet-American relations as a zero-sum game, in which any improvement of Sino-Soviet relations would automatically have an undesirable impact on Sino-American and/or Soviet-American relations. (Incidentally, Soviet officials shared a similar view of improvements in Sino-American relations.) Such a view probably peaked in the period 1978–81 when Washington, worried about its own relative weakness and captivated by the supposed strategic weight of the PRC in the global balance of power, envisioned a kind of quasi-alliance between the United States and the PRC to contain the USSR. China's proclamation of an "independent foreign policy" in 1982 and the resurfacing of Sino-American frictions over Taiwan arms sales took the wind out of the sails of this strategic fantasy.

The deliberate pace of Sino-Soviet amelioration, continuing clashes of interest over regional political issues (especially Cambodia and Afghanistan), and the Chinese leaders' assertions that political relations will not become normalized until a solution is reached to the Three Obstacles, all have had the effect of allaying American anxieties over Sino-Soviet relations. The United States' enhanced strategic position, the resumed forward movement of

U.S.-China relations (including in the military-security sphere), and Beijing's greater dependency on Western technology, credits, and investment further reinforce a relatively relaxed American mood. Although further substantial improvements in Sino-Soviet relations cannot be ruled out, they are also not very likely. Elements of rivalry and competition seem to be built into the relationship between the two Communist neighbors and will balance the amelioration in other areas that we have discussed. In Washington as in Moscow and Beijing there is at least agreement on one point—there is no going back to the Sino-Soviet alliance of the 1950s.

In conclusion, although the estrangement between China and the Soviet Union may have ended, it would be naive to expect anything approaching accord on all issues between two large contiguous powers with such different histories and divergent national interests. Vice Prime Minister Li Peng's remark to a group of American journalists made the point: "We hope that China and the Soviet Union will become good neighbors, but they will not become allies." Originally, China linked the normalization of bilateral relations with the Soviet Union to changes in Soviet foreign policy (Afghanistan, Cambodia)—a link that Moscow rejected. Yet the steady improvement in Sino-Soviet relations has occurred despite the Kremlin's position. It may be concluded, then, that the movement toward normalized relations has occurred largely on Soviet terms.

The Chinese, for reasons of their own, have welcomed progress in Sino-Soviet relations; at the same time they want to avoid the appearance of caving in to Moscow or withdrawing their demands for linkage. Although the Chinese assert that no progress has been made in the political relations between the two countries, this claim cannot be taken seriously. To accept it would be to say that trade, economic assistance, cultural exchanges, and tourism have no political meaning, to say nothing of high-level meetings and official visits.

China's dependence on the USSR as an unequal partner in the Sino-Soviet alliance of the 1950s was fundamentally out of character with China's history as an independent great power and with its aspirations to regain that status in the modern world. Similarly, the Sino-Soviet enmity of the 1960s and 1970s reflected a concatenation of factors that have either disappeared or lost their potency. At present, the Sino-Soviet relationship exhibits a combination of cooperative and conflictual elements in a balance that neither side has compelling reasons to upset. In this sense, the normalization of Sino-Soviet relations may be said to have occurred—and without producing any cataclysmic changes in the world balance of power or adversely affecting the interests of the United States.

Notes

1. Foreign Broadcast Information Service (cited hereafter as FBIS), *Daily Report: USSR*, 29 July 1986, p. R 14.
2. *People's Daily*, 8 September 1986, p. 1.
3. *Liaowang*, 29 July 1985, p. 27.
4. Steven I. Levine, "Some Thoughts on Sino-Soviet Relations in the 1980s," *International Journal*, Vol. 34, No. 4, Autumn 1979, pp. 652–54.
5. FBIS, *Daily Report: USSR*, 24 March 1982, p. R 3.
6. FBIS, *Daily Report: China*, 28 December 1984.
7. *China Daily*, 23 January 1986, p. 1.
8. Roy Medvedev, *China and the Superpowers* (New York: Basil Blackwell, 1986), p. 220.
9. *Pravda*, 19 March 1986, p. 4; *China Daily*, 19 March 1986, p. 1.
10. *International Affairs* (Moscow), 4 (April 1985), p. 9.
11. FBIS, *Daily Report: China*, 12 March 1985.
12. *Pravda*, 25 February 1986, p. 11.
13. Medvedev, *op. cit.*, p. 172.
14. *Beijing Review*, Vol. 29, No. 2 (13 January 1986), p. 7.
15. *Beijing Review*, Vol. 29, No. 4 (27 January 1986), p. 5.
16. *The Military Balance, 1985–1986* (London: International Institute for Strategic Studies, 1985), pp. 21–30.
17. Alfred Wilhelm, "National Security: The Chinese Perspective," in U. Alexis Johnson, et al., eds., *China Policy for the Next Decade* (Cambridge, Mass: Oelgeschlager, Gunn and Hain, 1984), pp. 181–219.
18. FBIS, *Daily Report: USSR*, 29 July 1986, p. R 14.
19. FBIS, *Daily Report: China*, 15 August 1986, p. C 1.

3. Japan, the Soviet Union, and the Northern Territories: Prospects for Accommodation*

Peggy L. Falkenheim

THE recent changes in the Soviet leadership and the more sophisticated foreign policy tactics of Mikhail Gorbachev have led to speculation among some observers of East Asian international relations that the Soviet Union may adopt a more flexible line in its territorial dispute with Japan in order to effect a radical improvement in Soviet-Japanese relations. According to this view, the Soviet Union has little to lose and much to gain by making concessions in its dispute with Japan over three islands and a small archipelago located east of Hokkaido. Moreover, the intransigent Soviet attitude on this issue has hurt Soviet interests by pushing Japan into the arms of China and by dissuading the Japanese from full participation in Siberian development and other economic relations with the USSR. These observers believe that these islands do not have sufficient strategic and economic importance to be worth this sacrifice. They further suggest that Soviet adoption of a more flexible attitude toward this issue would have an important positive impact on other aspects of Soviet-Japanese relations.

Events since Gorbachev's March 1985 assumption of power have suggested that Moscow is taking a greater interest in Japan. The accession of a new Soviet leadership has brought some

---

*Funding for part of this research was provided by the Donner Canadian Foundation.

change in Soviet tactics, including the adoption of a more positive tone in Soviet statements on Japan and a new willingness to allow top Soviet leaders to visit Japan. In contrast to his predecessor as foreign minister, Andrei Gromyko, who had stayed away from Japan for fear of being pressed on the northern territorial dispute, Eduard Shevardnadze went to Japan in mid-January 1986, the first visit by a Soviet foreign minister in ten years. Subsequently, Moscow agreed to a future visit by General Secretary Mikhail Gorbachev. If this visit takes place, it will be the first ever by a top Soviet leader. These visits and Moscow's appointment in June 1986 of a Japanese-speaking career diplomat as ambassador to Tokyo are important symbolically to the Japanese, who over the years have felt slighted by Moscow's failure to accord sufficient importance to their country.

Moscow's growing interest in Japan has increased the incentive for compromise on the territorial dispute. However, so far, the changes in Moscow's position have been purely tactical. There continues to be a wide gap between the Soviet and Japanese substantive positions on the Northern Territories. This gap will be difficult to overcome despite growing incentives for compromise. This chapter will describe the history of the dispute in order to explain why the Soviet and Japanese positions have grown further apart, not closer, over the years, and then evaluate the current incentives for and obstacles to its resolution.

I. Background to the Territorial Dispute

The Soviet-Japanese territorial dispute has a long history, dating at least as far back as the latter part of the seventeenth century when explorers and merchants from both Russia and Japan visited the Kurile Islands, located between the northernmost Japanese island of Hokkaido and the USSR's Kamchatka Peninsula. During this same period, explorers and settlers from the two countries also visited and colonized Sakhalin Island so that it too became an object of dispute between Russia and Japan.[1]

The first efforts to settle this territorial dispute were made in the mid-nineteenth century, when the Shogunate was pressured by Russia into establishing consular and trade relations. By the terms of the Treaty of Shimoda, which Vice-Admiral Evgenii

Putiatin negotiated with Japan in 1855, the southern Kurile Islands, specifically Etorofu and the islands to the south of it, were declared to be Japanese, and the northern Kuriles, Uruppu and the islands to the north of it, were assigned to Russia. A compromise formula was adopted regarding Sakhalin, by which the island was declared to be a joint possession of both countries.[2] The ambiguity regarding Sakhalin's status was resolved by the Sakhalin–Kurile Islands Exchange Treaty of May 7, 1875, by which Japan agreed to renounce its claims to Sakhalin in return for which Russia ceded the northern Kuriles to Japan.[3] This territorial settlement was modified by the Treaty of Portsmouth of September 5, 1905, ending the Russo-Japanese War, which allowed Japan to retain the southern half of Sakhalin, seized during the hostilities, but forced it to return the northern half to Russia.[4] During the Japanese intervention in Siberia after the Bolshevik Revolution, the northern half of Sakhalin was again occupied by Japanese troops, which were withdrawn as part of the settlement embodied in the Peking Convention of January 20, 1925, establishing diplomatic relations between Japan and the new Soviet regime.

This territorial division was radically altered at the end of World War II. At the February 1945 Yalta Conference, the Allies approved several conditions for eventual Soviet participation in the war in the Pacific, among them Soviet annexation of southern Sakhalin and of the islands adjacent to it and of the Kurile Islands. During the fighting which began on August 8, 1945, Soviet troops occupied these territories and the Habomai archipelago and Shikotan Island near Hokkaido.

The Red Army's occupation of these islands is the basis for the postwar Soviet-Japanese territorial dispute. Moscow claims that there is no territorial dispute between the Soviet Union and Japan because Soviet sovereignty over southern Sakhalin and the Kuriles was recognized by various Allied wartime agreements, by the terms of Japan's unconditional surrender, and by the 1951 San Francisco Peace Treaty in which Japan renounced its claims to these islands.[5] The Soviet position is that Kunashiri, Etorofu, Shikotan Island, and the Habomai archipelago, whose return is demanded by Tokyo, are all part of the Kurile Islands whose ownership Japan has renounced.

Moscow's contentions have been denied by Tokyo, which claims that it has not recognized Soviet sovereignty over southern Sakhalin and the Kurile Islands. Tokyo claims that it is not legally bound by the Yalta Agreement since its terms were still secret at the time Japan accepted the Potsdam Declaration as the basis for its World War II surrender. The Potsdam Declaration did, however, refer to the Cairo Declaration, which stated that Japan would be forced to renounce its claims to all territories taken by "violence and greed," a condition which does not apply to the Kuriles and southern Sakhalin.[6] Although Japan nevertheless (in the San Francisco Treaty) renounced its claims to these islands, the USSR was not a signatory to this agreement, which did not specify under whose jurisdiction they had passed. Therefore, Soviet sovereignty over southern Sakhalin and the Kuriles still has not been recognized by international agreement. Moreover, Japan maintains that Kunashiri, Etorofu, the Habomais, and Shikotan are not part of the Kurile chain whose ownership it renounced, that they have never belonged to a foreign power, and that they are inherent, inalienable Japanese territory. Tokyo cites historical, geographical, and botanical evidence to support this argument.[7]

While these arguments represent the two countries' recent stands on this issue, their positions had not hardened to this point in 1955 when they began peace treaty negotiations. At that time, there were strong incentives to reach agreement. The recently elected Japanese prime minister Hatoyama Ichiro was eager for Japan to conduct a foreign policy more independent of the United States. To do so, Japan wanted to gain admission to the United Nations, which was being blocked by a Soviet veto. Japan also wanted to end the state of war with the USSR and to reestablish diplomatic relations in order to arrange for the return of Japanese prisoners of war still being held in the Soviet Union. On the Soviet side, First Secretary Nikita Khrushchev was pursuing a policy of peaceful coexistence, attempting to resolve negotiable issues in East-West relations in order to free Moscow to focus its attention on making gains among developing countries. In support of this effort, Khrushchev demonstrated a certain flexibility, unusual in the Soviet context, regarding territorial questions and zones of influence when he agreed to give back the Porkkala

Naval Base to Finland and to relinquish the Soviet zone in Austria.

Because of these incentives for compromise, the Soviet Union and Japan came closer in 1955 to reaching agreement on the territorial question than they have since then. When the talks began, a wide gulf seemed to separate their positions on the territorial issue. The Japanese demand for the return of the Habomais, Shikotan, Kunashiri, and Etorofu was refused by the Soviet Union. But in August 1955, two months after the talks began, Moscow agreed to give back the Habomais and Shikotan. At this point, agreement seemed near since the Japanese delegate had come to London with the understanding that his government would accept such a compromise.[8]

However, a peace treaty embodying this compromise was not signed because Tokyo's territorial position suddenly hardened. Shortly after Moscow had made this concession, the Japanese Foreign Ministry published a pamphlet defending Japan's claim to Kunashiri and Etorofu. Then, a draft treaty was prepared in which the Foreign Ministry not only demanded the return of those two islands but even suggested that ownership of the northern Kuriles and southern Sakhalin should be decided by an international conference. Since this draft clearly was unacceptable to the Soviet Union, negotiations were broken off in late August.[9]

The hardening of Japan's negotiating posture has been attributed, in part, to pressure from Prime Minister Hatoyama's Liberal allies in the newly formed Liberal Democratic Party who opposed a territorial compromise. According to Shunichi Matsumoto, Japanese negotiator in these peace treaty talks, pressure from the United States also was very important. In order to forestall a Soviet-Japanese rapprochement, U.S. Secretary of State John Foster Dulles threatened to revise Washington's position that Okinawa, then occupied by the United States, was inherently Japanese territory if Tokyo recognized Kunashiri and Etorofu as Soviet territory.[10]

After several fits and starts, the deadlock finally was broken in October 1956 when the two sides decided to sign a Peace Declaration reestablishing diplomatic relations and ending their state of war. The USSR promised to return the Habomais and Shikotan after the conclusion of a future peace treaty resolving the territorial dispute.

By the time peace treaty talks were resumed in 1972, the Soviet and Japanese positions had grown much farther apart. In January 1960, the USSR backtracked on its pledge that it would return the Habomais and Shikotan after conclusion of a peace treaty, stating that the islands would not be returned until all foreign troops were evacuated from Japanese soil. This hardening of Moscow's posture was an effort to influence Japanese policy during negotiations for renewal of the U.S.-Japanese Security Treaty. In 1968, an across-the-board change in Soviet-published statements suggested that Moscow no longer intended to recognize even residual Japanese sovereignty over the Habomais and Shikotan.[11] By the end of the 1960s, Soviet leaders were justifying their refusal to make concessions on the territorial dispute with a new argument that it was wrong to tamper with the settlement made at the end of World War II. This new argument reflected Soviet concern about the impact of concessions to Japan on other powers, particularly China and West Germany.

During the late 1960s, Japanese began to pay increasing attention to the territorial dispute. In the preceding period, the northern territorial dispute was overshadowed by the Okinawa question, which was considered more important because many more Japanese lives had been lost defending it and because a large number of Japanese still lived there, whereas all the Japanese occupants of the Northern Territories had been evacuated at the end of World War II. As one reflection of Japan's national self-assertion in the 1960s, growing attention began to be paid to the Northern Territories. One sign of this change was the decision by the two houses of the Japanese Diet to send missions to Hokkaido in late August and September 1967 for an on-the-spot investigation of the northern territorial problem, the first mission of this kind.[12] In October of the same year, the ruling Liberal Democratic Party (LDP) created a special committee for the study of the territorial question.

Japanese efforts to promote the return of the Northern Territories increased after Prime Minister Sato's November 1967 trip to Washington when U.S. President Lyndon Johnson agreed to return the Bonin Islands and made a vague commitment to return

Okinawa. On August 28, 1968, the Japanese Prime Minister's Office drafted a "general outline of northern territory countermeasures," which stipulated that the Habomais, Shikotan, Kunashiri, and Etorofu were henceforth to be marked as Japanese on official government maps and provided for the appropriation of local taxes for Shikotan, Kunashiri, and Etorofu starting in fiscal 1969. (Taxes had been appropriated for the Habomai Islands since 1959.)[13] On October 21, 1970, Prime Minister Eisaku Sato explained Japan's territorial position to the United Nations General Assembly. These actions may have persuaded Soviet policymakers to adopt a harder position toward the Habomais and Shikotan because any signs of flexibility, rather than satisfying Tokyo, only encouraged it to make further territorial demands.

In the early 1970s, signs that the United States and Japan were moving closer to China induced the Soviet Union to reopen Soviet-Japanese peace treaty talks. Japan won a concession when Brezhnev signed a Joint Communiqué with Prime Minister Tanaka in October 1973 agreeing to continue negotiations for a peace treaty "resolving the yet unresolved problems remaining since World War II," an oblique reference to the territorial dispute. However, no substantive agreement was reached during these talks. It was reported that when Tanaka repeatedly raised the territorial issue, Brezhnev became "incensed" and finally "exploded in anger."[14]

Some analysts of Soviet-Japanese relations have argued that this outcome was not inevitable and that Japan missed an opportunity in the early 1970s to reach agreement with the Soviet Union. In their view, Moscow's desire to forestall a Sino-Japanese rapprochement and its growing interest in Siberian development cooperation created strong incentives for compromise. Japan reduced the incentives for compromise by moving so quickly to establish diplomatic relations with Beijing and by not making Japanese-Siberian development cooperation conditional upon successful resolution of the territorial dispute.

These analysts exaggerate the prospects for Soviet-Japanese agreement on the territorial dispute by failing to recognize the wide gap between the two countries' positions. Soviet flexibility regarding the Habomais and Shikotan was and still is always possible. However, even if Japan had delayed its rapprochement with China, it is doubtful that Moscow would have agreed to return Kunashiri and Etorofu, given these islands' strategic importance (see below). These analysts also exaggerate the incentives for compromise. Despite the signs of an improvement in American and Japanese relations with China in the late sixties and beginning of the seventies, there were a number of reasons for Moscow to feel confident that international trends were moving in a favorable direction. Moscow's achievement of strategic parity with the United States, the growing detente with Washington, increasing American disillusionment with its military role in Southeast Asia, growing tensions between the United States and Japan over trade and other issues, and Soviet success in persuading West Germany in 1970 to recognize the European territorial status quo were all seen in Moscow as positive developments. Soviet analysts also perceived a number of obstacles to American and Japanese efforts to move closer to China.

During the period after Tanaka's visit, there were even fewer incentives for Moscow to compromise. The 1973 oil shock reinforced the Soviet belief that territorial concessions were not needed to entice the Japanese into Siberian resource development cooperation. When Japan showed no sign of flexibility toward the territorial dispute, the Soviet Union again changed its tactics. During Japanese Foreign Minister Miyazawa's visit to Moscow in January 1975, the Soviet Union proposed the conclusion of a treaty of "good neighborliness and cooperation" as an interim measure while a peace treaty was being negotiated. However, Japan refused to sign this treaty because it bypassed the territorial dispute.

In the succeeding period, Soviet leaders adopted an even harder posture by denying that any territorial dispute existed. At the February 1976 Soviet Party Congress, Brezhnev called the Japanese position an "illegal territorial demand." In the same year, Moscow for the first time required Japanese visiting their relatives' graves on the Northern Territories to have valid passports and visas, a condition unacceptable to Tokyo.

The territorial conflict was exacerbated by Moscow's announcement on December 10, 1976, of its intention to establish a 200-mile economic zone encompassing the waters around the disputed islands. Even before this announcement, the territorial dispute had caused problems for Japanese fishermen. Throughout the postwar period, there were frequent Soviet arrests of Japanese fishermen caught operating in the territorial seas around the disputed islands. The Soviet Union and Japan had tried to resolve this problem by discussing an agreement guaranteeing safe operations for Japanese fishermen. However, the territorial dispute impeded agreement on the terms.[15] Moscow's declared intention to establish a 200-mile zone threatened to make this problem worse by greatly increasing the disputed sea area.

In order to resolve this problem, the Soviet Union and Japan entered into negotiations to establish rules governing Japanese fishing in the Soviet zone. They took place at a time when Soviet-Japanese relations already were strained by Tokyo's decision to allow the United States to inspect the MiG-25, landed on Hokkaido in September 1976 by a defecting Soviet pilot. They were complicated by Japanese concern that any agreement acknowledging Soviet jurisdiction over Japanese fishermen in waters surrounding the disputed islands would weaken Japanese territorial claims. After lengthy and rather bitter negotiations, the two sides agreed in May 1977 to a treaty including an article stating that no provision in this agreement "can be construed as to prejudice the positions . . . of either Government . . . in regard to various problems in mutual relations,"[16] an oblique reference to the territorial dispute. Japan also insisted that the clause defining the sea area for the treaty should refer to the Supreme Soviet Presidium's declaration of December 10, 1976, regarding the Soviet 200-mile zone rather than to a subsequent Council of Ministers' decision, because the former was less specific.[17]

During the next phase of the negotiations, which began in late June 1977 and was concluded on August 4, the two governments discussed an interim agreement to regulate Soviet fishing in the Japanese 200-mile zone established on July 1, 1977. One obstacle was that the Japanese 200-mile zone includes the waters around the disputed islands, overlapping the Soviet zone. But the two sides were able to work out a compromise formula that allowed them to sidestep the territorial problem.[18] Despite this outcome, Moscow's initial tough posture in the negotiations created lasting resentment in Japan.

On account of its uncompromising attitude, the Soviet Union may have missed an opportunity to keep Japan from moving closer to China. In February 1978, Japan signed a large-scale, long-term economic cooperation pact with China, and in August 1978 a peace and friendship treaty, abandoning a policy of equidistance between the Soviet Union and China. A more accommodating Soviet posture on the territorial dispute might at least have counterbalanced this improvement in Sino-Japanese relations, if not prevented it.

The prospects for resolving the territorial dispute became even dimmer in the period after conclusion of the peace and friendship treaty, which saw a marked deterioration of Soviet-Japanese relations caused by the Sino-Japanese rapprochement, the Soviet military buildup in the Asia-Pacific region and on the disputed islands, the Soviet invasion of Afghanistan, the KAL 007 tragedy, and frosty U.S.-Soviet relations. The Soviet Union reacted to the reopening of Sino-Japanese talks by increasing the pressure on Japan to sign a good neighborliness and cooperation treaty. When the Soviet government organ *Izvestia* published a draft treaty in late February 1978, Foreign Minister Sonoda proclaimed that Japan would not consider it until after the conclusion of a peace treaty returning the Northern Territories.[19]

Soviet leaders also adopted a harder line regarding the Habomais and Shikotan. When Kono Yohei, leader of the New Liberal Club, visited Moscow in November 1978, he was told by Kosygin and other Soviet leaders that the Soviet Union was no longer bound by its 1956 pledge to return these islands when a peace treaty was signed. Kosygin reminded Kono of the Soviet declaration issued in 1960 proclaiming that these islands would not be returned until all foreign troops were withdrawn from Japan.[20] The same point was made by Soviet Ambassador Dimitri Polyansky, who also told a *Mainichi* correspondent that "the Soviet Union has no intention of transferring to Japan a single piece of stone, let alone an island."[21] These statements caused some consternation in Japan since the Soviet position on this issue had been ambiguous, and Tokyo has denied that Moscow has the right to alter unilaterally the terms of the 1956 pledge.

In late January 1979 the Japanese Defense Agency announced that the Soviet Union had reinforced its garrisons and bases on Kunashiri and Etorofu, two of the disputed islands. In a buildup which began in May 1978, the Soviet Union had increased the level of forces from around 2,000 to 5,000, and extended runways, improved port facilities, constructed new buildings and radar stations, and deployed surface-to-air missiles there.[22] In reaction, Japanese Deputy Foreign Minister Takashima Masuo delivered a verbal protest to Soviet Ambassador Polyansky in early February demanding the removal of Soviet troops and bases from the islands. Polyansky rejected Japan's protest, claiming that it constituted unwarranted interference in Soviet internal affairs since the islands were Soviet territory.[23] Subsequently, in February 1979, the Japanese Diet passed a resolution proclaiming these islands as Japanese territory and urging their government to demand the immediate withdrawal of Soviet military forces. When the Japanese Ambassador in Moscow, Uomoto Tokichiro, tried to present a copy of this resolution to Soviet Deputy Foreign Minister Nikolai P. Firyubin, the latter refused to accept it, declaring that this issue "may be . . . of great concern for the Japanese but it is not for the Russians,"[24] a remark sure to offend Japanese sensibilities.

Despite Japan's protests, the Soviet Union continued to increase its forces on the disputed islands. By the summer of 1979, the number of Soviet troops stationed on Kunashiri and Etorofu had been increased from a brigade level of 3,000 to 4,000 to a division level of 10,000 to 12,000 armed with attack helicopters, tanks, and heavy artillery, which suggested that their mission was not purely defensive. In the summer of 1979, regular Soviet ground forces were stationed for the first time on Shikotan Island.[25]

The Afghanistan invasion had a significant impact on Japanese security perceptions, creating growing concern about a prospective Soviet threat. This concern was reinforced by the continued Soviet military buildup during the 1980s in the Asia-Pacific region and on the northern islands. By late 1982, the number of Soviet troops on the disputed islands was increased to 14,000, and the subsonic MiG-17s on Etorofu had been replaced by supersonic MiG-21s.[26] In September 1983, twenty MiG-23 fighters were deployed on Etorofu. By April 1984, their number was increased to forty.[27]

The sanctions imposed after the Afghanistan invasion also produced a marked reduction in official Soviet-Japanese political contacts. Even before the Afghanistan invasion, Soviet Foreign Minister Gromyko had repeatedly postponed a planned visit to Japan to attend the next scheduled meeting between the foreign ministers of the two countries. Gromyko was reluctant to visit Japan because of fear that he would face pressure on the territorial issue. Formal high-level Soviet-Japanese political contacts were reduced even further in the first few years of the 1980s when periodic, regular consultations at the subministerial level were suspended as a consequence of Japanese sanctions after the Afghanistan invasion and the imposition of martial law in Poland. The absence of official contacts meant that no formal territorial negotiations were held during this period. In order to keep up the pressure on this issue, the Japanese took a variety of measures, among them visits by high-level officials to Cape Nosappu to inspect the northern islands, a visit by a Hokkaido delegation to New York to state Japan's case regarding the Northern Territories to various United Nations officials and ambassadors to the United Nations,[28] and the collection of 34 million Japanese signatures on an appeal for the return of the northern islands.[29] Whenever possible, Japanese raised the territorial issue with Soviet leaders, who reacted by denying that any territorial dispute existed.

The one exception to this negative trend was the fisheries area, where Soviet-Japanese interdependence created a need for regular contacts to discuss quotas and other issues. In August 1981, the two countries concluded an agreement to allow Japanese fishermen to collect sea tangle in the area around Kaigara Island in the Habomai chain. This agreement replaced a 1963 private agreement, abrogated in 1976 by the Soviet Union. The Soviet Union made some concessions during negotiations for this agreement by dropping its demands that the agreement use the Soviet name of the island instead of the Japanese name and that Japanese fishermen be required to have licenses issued by the USSR and fall under Soviet jurisdiction. Tokyo found these demands unacceptable because they could be interpreted in a way which would support Moscow's position in the territorial dispute. In return for these concessions, Japanese fishermen agreed to pay a substantial fee for the right to collect tangle. In December 1984, the Soviet Union and Japan concluded a new three-year pact regulating Japanese fishing in the Soviet 200-mile zone and Japanese fishing in the Soviet zone. This new pact replaced the two separate one-year agreements concluded in 1977 and renewed annually since then. In contrast to the prolonged and bitter negotiations in 1977, this time the territorial dispute did not impede agreement on fisheries' regulations.

Starting in 1984, there was some improvement in overall Soviet-Japanese relations as the two sides reestablished formal high-level political contacts suspended since the Afghanistan invasion. The years 1984 and 1985 saw an exchange of parlimentary delegations, and the reactivation of subministerial consultations about the Middle East, the United Nations, arms control, and other matters. However, Gromyko repeatedly refused invitations to visit Japan, saying that he would come only when it could be guaranteed that the discussion would not be dominated by the territorial dispute and that he would not face hostile demonstrations.

After Gorbachev's accession to power and Gromyko's replacement by Shevardnadze as foreign minister, the Soviet Union agreed to an exchange of visits by their foreign ministers for the first time since the late 1970s and to a visit to Japan by General Secretary Gorbachev. These decisions reflected a greater emphasis by the new Soviet leaders on Asia and on courting U.S. allies in contrast to their predecessors' primary focus on the U.S.-Soviet relationship. During his January 1986 visit to Tokyo, Shevardnadze adopted a style which the Japanese referred to as "smile diplomacy," smiling and joking in contrast to past Soviet leaders who often had offended the Japanese by making overt threats. During this visit and Abe's May 1986 visit to Moscow, the two foreign ministers reached agreement on a number of relatively noncontroversial issues, among them the resumption of scientific and technological cooperation, suspended since the imposition of martial law in Poland, a new five-year trade and payments agreement, a new costal trade treaty, a treaty for avoidance of double taxation, and a cultural exchange agreement. They also agreed to resume regular foreign ministerial exchanges and exchanged invitations for visits by their top leaders.

Regarding the territorial dispute, however, the positions of the two countries have remained far apart, although there has been some change in Soviet tactics. During his visit to Tokyo, Shevardnadze exhibited some flexibility by agreeing to resume peace treaty talks and by discussing the territorial dispute for three hours, instead of dismissing it with a statement that no such dispute exists. Following a pattern established during foreign ministerial visits in 1975 and 1976, the Joint Communiqué issued at the end of Shevardnadze's visit referred to the 1973 Tanaka-Brezhnev Joint Communiqué in which the two sides had agreed to continue peace treaty negotiations that would include consideration of the territorial dispute.[30]

The Soviet Union also has agreed for the first time since 1975 to allow Japanese to visit their relatives' graves on the Northern Territories without requiring them to have valid passports and visas. The first sign of Soviet flexibility toward this question was a statement in the Abe-Shevardnadze Joint Communiqué that the Soviet Union promised to consider "with all due attention from the humanitarian standpoint" Japan's request for such visits. When Abe visited Moscow in May, Gorbachev agreed to allow Japanese without visas to visit their relatives' graves on the Northern Territories on the condition that Soviet citizens were allowed to visit family graves in Japan on the same basis.[31] This proposal was unacceptable to Japan's Foreign Ministry, which feared that it would undermine its territorial position. Tokyo insisted that if Soviet citizens without visas were allowed to visit graves in Japan, then Japanese without visas should be allowed to visit graves on the mainland Soviet Union.

At the beginning of July, the two sides reached agreement and exchanged documents allowing Japanese without visas to visit their relatives' graves on the Northern Territories, Sakhalin, and several places on the Soviet mainland, and Soviet citizens without visas to visit their relatives' graves in Japan. The documents stated that such visits do not prejudice the two sides' legal positions on other questions, an oblique reference to the territorial dispute. They also specified that the sites for such visits will be determined annually. This will give the Soviet side continuing leverage over Japan on this issue since these visits are much more important to the Japanese involved than to the Soviet citizens.[32]

Despite these changes in tactics, the overall Soviet position on the territorial dispute has remained unyielding. During a news conference at the end of his January 1986 visit to Japan, Shevardnadze emphasized that there has been no change in Moscow's territorial position and urged Tokyo to adopt a more "realistic" posture.[33] During Abe's May visit to Moscow, Gorbachev said that there was no territorial dispute between the Soviet Union and Japan and criticized Japan for raising this question.[34] Tokyo's position has remained equally inflexible. The Japanese government not only has refused to make concessions on the territorial dispute but has tried to pressure the Soviet Union into making concessions by refusing to sign a long-term economic cooperation agreement until the territorial dispute is resolved.[35]

II. Obstacles to and Incentives for Resolution of the Dispute

This history of the northern territorial dispute has shown that the recent positions of the Soviet Union and Japan are more rigid and farther apart than they were thirty years ago. For the Japanese, the islands' symbolic importance has increased over the years as a reflection of growing national self-assertion and Japan's increasingly active and independent international role. The government's position that the northern islands are inherent, inalienable Japanese possessions has received increasing support from public opinion and from the opposition parties.

The Soviet Union has refused to make concessions on the territorial issue largely because of the islands' strategic significance, which has increased in the seventies and eighties as the Northern Territories have become an important support for the Soviet military buildup in the Asia-Pacific region and for Soviet efforts to maintain strategic parity with the United States. Two of the disputed islands, Kunashiri and Etorofu, are important to the USSR because of their strategic location. Possession of these islands allows Soviet armed forces to exercise greater control over the entrances and exits to the Sea of Okhotsk. Soviet submarines carrying intercontinental ballistic missiles capable of hitting the continental United States are based there in order to avoid the threat that sophisticated American antisubmarine warfare capabilities would pose to them if they were to roam the open seas. The Sea of Okhotsk is important as a logistical supply route for the Soviet naval base at Petropavlovsk on the Kamchatka Peninsula. Since the overland supply route to Petropavlovsk is very long, vulnerable, and difficult to travel during the winter, many of the supplies for the base there are transported by ship across the Sea of Okhotsk. Kunashiri and Etorofu also now have air bases used by Soviet planes reconnoitering Japan's Pacific Coast and electronic facilities used to monitor Japanese military communications.

Soviet possession of Etorofu increases naval access to the Pacific because ships and submarines stationed in its deep, ice-free harbor do not have to go through chokepoints controlled by foreign powers. Aerial surveillance of their movements often is impeded by fog, allowing them undetected access to the Pacific Ocean.[36] Possession of the islands also gives the USSR control over the Kunashiri Channel, one of the three main routes used by the Soviet Far Eastern Fleet to reach the Pacific, and the only one not under the control of the United States and its allies.

Soviet leaders further have worried about the military use that could be made of the islands if they were returned to Japan. One justification given for Soviet refusal to return the islands is that nothing would prevent Japan from allowing the United States to build military bases on them. In June 1982, then Foreign Minister Fukuda announced a concession designed to meet this objection, saying that Japan would agree to keep the Northern Territories demilitarized if they were returned to Japan,[37] but Moscow so far has not responded positively to this overture.

While strategic considerations have been the most important factor behind Soviet intransigence in the territorial dispute, economic considerations are not negligible. The Northern Territories are surrounded by one of the world's three richest fishing grounds, which has become increasingly important now that the Soviet catch in other regions has been reduced by the establishment of 200-mile economic zones. Another factor behind Soviet intransigence is a fear of the effect that concessions would have on territorial demands against the Soviet Union by other powers, particularly China.

Although the Soviet and Japanese positions on the territorial dispute have become more rigid over time, there are some incentives for compromise. One incentive is the desire of both sides to increase economic interchange. Soviet and Japanese economic interests are complementary. Moscow is eager to obtain access to Japanese high technology and Japanese assistance in developing light and consumer goods industries, which are slated to play an increasingly important role in the future Soviet economy. The Soviet Union needs Japanese assistance in the development of the natural resources located in Siberia around the new Baikal-Amur Mainline (BAM) railroad. The huge Soviet investment in this railroad will not pay off unless it can be used to export newly developed Soviet resources to Japan and other Pacific countries. For the Soviet Far East, Japan is an important export market and a supplier of machinery, equipment, and consumer goods used to raise productivity in local industry and to supply the local population more quickly and efficiently than if goods had to be imported from distant regions of the USSR.[38]

Despite this complementarity, Soviet-Japanese trade still is only a small percentage of each country's total foreign trade. In the early 1980s, the level declined largely for economic reasons—in particular, decreasing Japanese demand, at least in the short run, for Soviet natural resources, the high value of the yen which has made Japanese goods and services more expensive than those of Japan's West European competitors, and a marked trade imbalance in Japan's favor. Political factors, in particular, the post-Afghanistan and Polish sanctions and Cocom restrictions on the export of high technology to the Soviet Union, also acted as a constraint. Recently, there was a slight upturn in the level of Soviet-Japanese trade, but its impact was limited by sharply declining prices for oil, natural gas, and gold, three principal Soviet exports.

Not only has trade not been great enough to serve as an inducement for the Soviet Union or Japan to make territorial concessions, but each side has tried, without success, to use the other's desire for increased trade as a means of pressuring it to adopt a more flexible position. For Japan, the Soviet Union still is not an important enough current or prospective economic partner for large corporations with political clout to pressure the government into making concessions on the Northern Territories. For the Soviet Union, increased trade with Japan could help improve Soviet economic performance and productivity, two important Gorbachev objectives. However, the Soviet Union often can obtain the same kinds of assistance from West Germany, France, and other West European countries as it can from Japan. This reduces the incentive for Moscow to make territorial concessions in order to promote Soviet-Japanese economic cooperation.

Pressure for compromise comes from Japanese fishermen who operate in Soviet-controlled waters around the disputed northern islands, an area rich in crab, sea urchins, abalone, and other valuable marine products. Access to these waters has become increasingly important to Japan since the creation of 200-mile economic zones in the mid-1970s expanded the ocean area around the Northern Territories claimed by Moscow and reduced Japanese fishing operations in other regions. While some Japanese fishermen are strong supporters of Japan's irredentist demands, most are more concerned about gaining access to Soviet-controlled waters than about pressing Japan's territorial claims. The Soviet Union has tried to take advantage of their concerns to undermine support for Japan's irredentist movement, which is particularly active in Hokkaido, home of the former residents of the Northern Territories and of most of these fishermen. Fishermen who join the Soviet-Japan Friendship Association have been accorded preferential treatment by Soviet patrol ships. Special treatment also has been accorded to Japanese fishermen who agree to spy for the Soviet Union by reporting on the activities of Japanese defense forces in Hokkaido and of the irredentist movement.[39] So far, however, these Soviet efforts have not produced useful leverage over Tokyo's territorial policy. Soviet attempts to build up a strong friendship movement in Hokkaido have met with strong resistance from irredentist groups, who recently forced the closing of an unused Soviet-Japanese friendship hall. Soviet efforts to use Japanese fisheries interests to gain leverage have floundered because Japan can exert counterleverage by controlling Soviet access to the Japanese 200-mile zone. Moreover, Tokyo has demonstrated that it is willing, if necessary, to sacrifice Japanese fisheries interests in order not to undermine its territorial position.

A growing incentive for Soviet flexibility is the negative effect that its intransigence has on Japanese security policies and perceptions. The Soviet Union has adopted an inflexible position toward the Northern Territories largely because of their strategic value. However, Soviet intransigence, while protecting a valuable piece of real estate, is undermining Soviet security in other ways through its effect on Japanese attitudes and perceptions. In the past, Japan looked for a Soviet concession on the territorial dispute as a token of Soviet good intentions and confirmation of the correctness of Japan's omnidirectional foreign policy of guaranteeing security by maintaining friendly relations with all countries. This policy has been a success with other countries, but not with the Soviet Union.[40] Moscow's tough stand on the Northern Territories, the Soviet military buildup in the Asia-Pacific region, and the militarization of the northern islands, have all produced an increased Japanese sense of threat and have induced Tokyo to place greater reliance on military strength and its alliance with the United States as guarantors of Japanese security. This attitude was manifest at the sixteenth meeting of the U.S.-Japan bilateral security committee, which was held in Honolulu at the same time that Shevardnadze was visiting Japan. At the meeting in Hawaii, which focused on the growing Soviet military threat to the Asia-Pacific region, Japanese Deputy Foreign Minister Yanai Shinichi told the gathering that Japan was under no illusions regarding the Soviet Union.[41]

From the perspective of the leaders of both countries, the territorial stalemate has impeded the achievement of important objectives. As part of Gorbachev's new "peace offensive" in the Asia-Pacific region, noted by his July 1986 Vladivostok speech, there have been renewed Soviet efforts recently to gain support for the convening of a Pan Asian Security Conference and the adoption of confidence-building measures in that region. Despite Japanese pacifist sentiments, Soviet leaders have tried repeatedly, but without success to date, to obtain Tokyo's support for these initiatives, which are viewed with suspicion in part because they might imply acceptance of the territorial status quo. On the Japanese side, Prime Minister Nakasone is eager to reach an accommodation with Moscow before he leaves office. He has placed a particular emphasis on foreign policy, and relations with the Soviet Union are the main area where his foreign policy has not been particularly successful. Nakasone has been pressing for a summit meeting with Gorbachev. At one point, Gorbachev was expected to visit Tokyo in early 1987 but his visit has been postponed.

Given Gorbachev's demonstrated willingness to alter Soviet

policies on other issues, there has been speculation that if and when he visits Japan, he may propose some change in Soviet policy toward the Northern Territories. While this is conceivable, there still are significant barriers to a territorial settlement. It is highly unlikely that Gorbachev would agree to return Kunashiri and Etorofu, given these islands' strategic importance. He might be willing to return the Habomais and Shikotan but this offer would meet strong resistance in Tokyo if Gorbachev made their return conditional upon Japan's renunciation of any further territorial claims. For Gorbachev to offer to return the Habomais and Shikotan, without such a condition, would represent a radical change in Soviet policy toward the Northern Territories. So far, there is no indication that Gorbachev favors such a change. Even if he did, he would face significant Soviet resistance. A compromise along these lines might be accepted by Japan but not without some significant opposition. The opposition parties and even some supporters of the Liberal Democratic Party have shown a willingness to accept a territorial settlement that would provide for the return of the Habomais and Shikotan but would postpone consideration of the question of Kunashiri and Etorofu until a later date, when improved Soviet-Japanese relations would create more favorable conditions for its resolution. A settlement along these lines might appeal to Nakasone but would meet strong resistance from Soviet specialists in Japan's Foreign Ministry. Japan might be more willing to accept a compromise in which the Soviet Union signed a peace treaty returning the Habomais and Shikotan and recognizing residual Japanese sovereignty over Kunashiri and Etorofu, while retaining control of them. However, such an offer would represent a radical departure from previous Soviet policy, one that seems improbable given the strategic importance of Kunashiri and Etorofu.

In view of these obstacles, a territorial settlement seems unlikely. If and when Secretary General Gorbachev visits Japan, he may try to improve relations by offering to reduce Soviet troops on the northern islands. Reportedly, such an offer has been discussed in the preliminary negotiations for Gorbachev's visit, but the Soviet side evidently has made a troop reduction conditional upon Tokyo's willingness to impose limitations on U.S. forces in Japan. This condition has been rejected by Tokyo. If Moscow were to reduce Soviet troops on the northern islands without imposing unacceptable conditions, this could decrease the strains in Soviet-Japanese relations, but it would not resolve the territorial dispute. A territorial settlement seems unlikely without a radical reduction in East-West tensions, both globally and in the Asia-Pacific region, and a major shift in emphasis from military to nonmilitary means of guaranteeing Soviet security.

Notes

1. John A. Harrison, *Japan's Northern Frontier* (Gainesville, Fla: University of Florida Press, 1953), pp. 12–38.
2. George Alexander Lensen, *The Russian Push Toward Japan: Russo-Japanese Relations, 1697–1875* (Princeton, N.J.: Princeton University Press, 1959), pp. 331–337.
3. Harrison, *Japan's Northern Frontier*, pp. 55–56.
4. Eugene P. Trani, *The Treaty of Portsmouth: An Adventure in Diplomacy* (Lexington, Ky.: University of Kentucky Press, 1969), pp. 96–97, 145–150.
5. See, for example, D. Petrov. "Development of Soviet-Japanese Relations," *International Affairs* (Moscow) (September 1969), p. 34; TASS broadcast, 1 December 1969.
6. As the preceding discussion has shown, Tokyo was correct when it claimed that the Kurile Islands had not been taken by "violence and greed," since Japan acquired them through the Sakhalin-Kurile Island Exchange Treaty of 1875. However, the same was not true of southern Sakhalin, which Japan seized during the Russo-Japanese War.
7. See, for example, Japan: Ministry of Foreign Affairs, *Japan's Northern Territories*, 1980.
8. "Interview with Matsumoto Shunichi [Japan's delegate to the London negotiations], January 20, 1963," cited by Donald C. Hellmann, *Japanese Foreign Policy and Domestic Politics: The Peace Agreement with the Soviet Union* (Berkeley: University of California Press, 1969), p. 34.
9. *Ibid.*, pp. 34–35.
10. Shun-ichi Matsumoto, *Northern Territories and Russo-Japanese Relations* (Sapporo, Hokkaido: Japan League for the Return of the Northern Territories, 1977), pp. 13–14.
11. For a more detailed discussion of this point, see Peggy L. Falkenheim, "Continuity and Change on Soviet Policy Toward

Japan, 1964 to 1969" (Ph.D. dissertation, Columbia University, New York, 1975), pp. 120–124.
12. "Northern Territory in the Dim Light," *Asahi*, 10 September 1967, trans. in *Daily Summary of the Japanese Press* (cited hereafter as *DSJP*). 14 September 1967, pp. 9–11.
13. "LDP Northern Territorial Problem Special Committee to Hold First General Meeting on October 13," *Yomiuri*, 7 October 1967, trans. in *DSJP*, 10–11 October 1967, p. 25; "Northern Territory to Be Treated as Territory from Administration Aspects; Clarification in Maps; Distribution of Local Taxes," *Nihon Keizai*, trans. in *DSJP*, 29 August 1968, p. 18.
14. *Tokyo Shimbun*, 12 October 1973.
15. For more details, see Peggy L. Falkenheim, "Some Determining Factors in Soviet-Japanese Relations, *Pacific Affairs*, 50:4 (Winter 1977–78), pp. 608–613.
16. *Asahi*, 25 May 1977, trans. in *DSJP*, 27 May 1977, pp. 5–8.
17. *Kyodo*, 25 May 1977.
18. *Daily Yomiuri*, 5 August 1977, p. 1.
19. *Kyodo*, 24 February 1978.
20. *Sankei*, 25 November 1978, p. 1 (in *DSJP*, 2–4 December 1978, p. 16); *Mainichi Daily News*, 24 November 1978, p. 1.
21. *Mainichi Daily News*, 25 November 1978, p. 1.
22. *New York Times*, 31 January 1979, p. A6. Since this information was known for some time to the JDA, it is not clear why it chose that particular time to announce it. One possibility is that it hoped to influence the budget debate then about to take place in the Diet.
23. "Editorial," *Japan Economic Journal*, 13 February 1979, p. 10.
24. *Mainichi Daily News*, 26 March 1979, p. 1.
25. *Japan Economic Journal*, 9 October 1979, p. 3; John Lewis, "Inadequate Bear Traps in the North," *Far Eastern Economic Review*, 14 March 1980, p. 23.
26. *New York Times*, 30 December 1982.
27. *Mainichi Daily News*, 28 April 1984, p. 1.
28. *Mainichi Daily News*, 17 November 1983, p. 2.
29. *Mainichi Daily News*, 19 January 1984, p. 16.
30. This reference was worded as follows: "The two foreign ministers conducted negotiations on the conclusion of a peace treaty, including the problems which might constitute the content of said treaty, on the basis of the agreement affirmed in the Joint Communiqué of October 10, 1973." The phrase "problems which might constitute the content of said treaty" is an oblique reference to the territorial dispute (Jiji Press News Service, 20 January 1986).

31. In several places in Japan there are graves of Russian prisoners captured during the 1904–05 Russo-Japanese War.
32. *Mainichi Daily News*, 13 June 1986, p. 12; *Japan Times*, 2 July 1986, p. 1, and 3 July 1986, p. 4.
33. *Mainichi Daily News*, 20 January 1986, p. 1; Jiji Press News Service, 20 January 1986.
34. *Mainichi Daily News*, 1 June 1986, p. 1.
35. Jiji Press News Service, 16 January 1986.
36. The Japanese Fleet assembled here in November 1941 before setting sail for Pearl Harbor.
37. Shojiro Arai, "A Special Report on Japan-Soviet Relations," *Business Japan* (June 1985), p. 24.
38. N. L. Shlyk, "Eksportnaya spetsializatsia Dal'nego Vostoka: osnovnye napravlenia razvitia (Export Specialization of the Far East: Main Directions of Development)," *Akademia nauk SSR, Izvestia Sibirskogo otdelenia: Seria Obshchestvennykh nauk*, 6 (May 1981), p. 35.
39. *Mainichi Daily News*, 28 January 1983, p. 12.
40. Hiroshi Kimura, "Japan-Soviet Relations: Framework, Developments Prospects," *Asian Survey*, Vol. 20, No. 7 (July 1980), pp. 710–712.
41. Jiji Press News Service, 16 January 1986.

4. Stability and Instability in the Sea of Japan*

Edward A. Olsen

THE Sea of Japan (Nihonkai) is a potentially troublesome body of water. Even its name is a subject of controversy because Koreans, regardless of their political persuasion, adamantly reject that name—preferring to call it the Eastern Sea (Dong Hae). While granting the Koreans the geographic logic of their nomenclature for a sea to their east, here we shall use the term "Sea of Japan" because it is internationally accepted.

What (aside from its name) makes the Sea of Japan potentially a trouble spot? Quite simply, it is one of the most strategic bodies of water on earth. Because it harbors the headquarters of the Soviet Union's Pacific Fleet (Moscow's largest), it represents a distant stronghold of the Soviet Eurasian "empire." Vladivostok (Ruler of the East) anchors the Soviet Union's claim to be a Pacific power. Via the Pacific Fleet's main ports at Petropavlovsk and Sovietskayagavan, Vladivostok shelters and sustains the USSR's power projection capabilities in the Pacific, and helps put substance behind Moscow's Asian diplomacy. Precisely because of those factors, both Japan—whose archipelago defines the sea which bears its name—and the United States—with great intrinsic interests in limiting Soviet gains and specific interests in seeing that its friends/allies (in Japan and South Korea) and its adversaries (North Korea) on the Sea do not fall under Soviet influence—are concerned about the stability of the Sea of Japan.

---

*Revision of an article printed in the Winter 86 issue of the *Journal of Northeast Asian Studies*. The original version was presented at the 1986 annual meeting of the International Studies Association. The views expressed are solely those of the author and do not necessarily reflect those of his employer.

In short, the Sea of Japan is a watery fulcrum around which the world's two military superpowers engage in part of their deadly global contest. It constitutes a major prize in that contest. The world's other superpower—in economic terms—Japan, is a reluctant player in this risky game. Observing this trio are two relatively small powers (North Korea and South Korea) which view the game being played out in their vicinity with great interest, but—remarkably—at arm's length. Just beyond the Japan Sea arena is the People's Republic of China (PRC), a rapt participant observer, trying to influence all parties.

In that context, this chapter will address a number of perspectives on Japan Sea security. The strategic posture of the four countries bordering the Sea of Japan plus the United States and PRC will be examined in relation to each other. Their politico-economic interests, threat perceptions, existing forces, and the alliance relationships that support one another's abilities will be assessed. Next, an evaluation will be made of each countries' strategic, political, and economic options and intentions that pertain to the sea. Finally, a set of policy recommendations will be offered with an eye on strengthening the positions of the United States and its allies toward the Japan Sea.

I. USSR

The Soviet Union has the most explicit set of strategic interests focusing on the Sea of Japan. At the southernmost portion of Soviet Far Eastern territory is the headquarters of the USSR's huge Pacific Fleet, near the northern end of the Japan Sea. It controls the nearby main ports for that fleet, Petropavlovsk on Kamchatka's Pacific coast and Sovietskayagavan across from Sakhalin. While this chapter's task is not an analysis of that fleet, one cannot understand the importance of the Sea of Japan to Moscow without reference to the fleet's capabilities and purposes. The table on page 72 outlines the USSR Pacific Fleet:

The Pacific Fleet has four basic purposes: to reaffirm the Soviet's open-ended presence as a Pacific power; to use that presence as means to exert influence over its Asian neighbors; to project its Asia-based forces outward; and to secure access to offshore sites for surface and underwater mobile launch platforms

TABLE I[1]
USSR PACIFIC OCEAN FLEET

Surface Combatants	441
Submarines	134 (31 SSBN/SSB and 103 General Purpose Boats)
Naval Aviation	500
Naval Infantry	1 Division

for "strategic" (i.e., nuclear[2]) warfare. Each of these purposes requires some elaboration.

Moscow's need to assert its presence reflects the still tenuous nature of its Far Eastern outposts. Inherited from a Czarist eastward-looking manifest destiny comparable in some ways to the American westward-moving frontier, they are thinly settled and far-flung enclaves of European civilization on the fringe of, and confined by, an inhospitable cultural and natural landscape. Given Great Russian psychological insecurities vis-à-vis a legacy of Asian (Mongol-Tartar) bogeymen, recurring anxieties about their European versus their Asian heritage, and the racism these factors have engendered among Russians, their precarious foothold in the far reaches of Eurasia instills in Soviet leaders a strong desire to reinforce their presence.

Seen against that background and the sense of implicit vulnerability the USSR displays, their assertiveness in showing the flag takes on added significance. General Secretary Mikhail Gorbachev's much ballyhooed July 28, 1986, speech on the USSR's Asia policy underlined this significance.[3] Moscow is intent upon establishing its credibility as a major entity in Asian affairs. The Kremlin now is exploring publicly a wider range of supposedly less coercive policy options, but behind them all still loom the ominous realities of the marked Soviet armed buildup in Asia. Projecting its forces and using them to influence Asian states is not a new tactic; all major powers have used such methods. Now, for the first time, the USSR has the wherewithal to flaunt its power further afield than it had previously. To the extent this reflects Soviet egos, it can be considered relatively benign. Similarly, to the extent it reflects a Soviet desire to proclaim its version of freedom of the seas, it is tolerable. However, there are two ominous facets of Soviet willingness to roam more widely than they have before. Because of a Soviet tendency to probe for weakness and take advantage of any weakness they come across by intimidating targets of opportunity, there are several states in Asia that are vulnerable to "Finlandization." Just as worrisome is the prospect of a Soviet version of containment encircling the edges of Eurasia via actively friendly states, nonaligned states, and intimidated states. That prospect gives rise to geopolitical visions of Moscow controlling Mackinder's "Heartland" *and* Spykman's "Rimland," using a version of Mahan's "Seapower" thesis.[4] Such conventional strategic moves may seem mundane in the last quarter of the twentieth century, but—if fulfilled—they could well adversely influence the balance among the powers which undergirds contemporary concepts of the nuclear armed "strategic" balance.

The last of Moscow's purposes is the easiest for observers of the international system to comprehend. Both superpowers seek diversity and dispersed basing for their nuclear forces in order to increase the uncertainties that provide the element each hopes will prevent the other from committing national suicide and dragging virtually all of humanity down with them. Soviet desires for such mobility are hindered greatly by climatic factors. A paucity of ice-free ports is nearly as troublesome for the Commissars as for the Czars. Soviet problems of this sort in the Northeast Asia/Northwest Pacific region are very comparable to their problems in the Barents and Norwegian seas. The USSR position and maneuverability in the Sea of Japan and Sea of Okhotsk is similarly constrained by weather and adversaries. Why, then, do most defense analysts seem more concerned about offshore Europe than offshore Asia? Once again, it seems to be mainly because of an old-fashioned factor in international relations: physical geography.

Unlike offshore Europe, where Soviet forces are constrained and contained by an opposing array of forces, in the Sea of Japan/Sea of Okhotsk combined region Soviet naval forces are constrained by the natural blockade formed by both the Japanese and Soviet sovereign- and Soviet-occupied (but Japanese-claimed) islands strung out like a chain in the enveloping waters of the Soviet Far East north of Japan. Those Soviet islands are separated by channels that are too shallow and/or vulnerable to ice blockage to be relied upon year round. Least useful is the shallow Tatar Strait, Moscow's only Soviet-controlled waterway between the Seas of Japan and Okhotsk. This strait frequently freezes solid. The most usable straits in the region are the ones between Japanese islands (especially the *tsugaru kaikyo* between Hokkaido and Honshu), the *tsushima kaikyo* and Korea Strait between Japan and South Korea, and the *soya kaikyo* between Hokkaido and Sakhalin. Neither Soya, which is icy and difficult in winter, nor Tsugaru, which is twisting and has fast currents, can be relied upon. Moreover, none of these waterways is easily controllable by Soviet forces because of the disposition of its adversaries' forces. This is especially true of the only decent point of egress, between Japan and Korea, where Soviet forces would have to run a severe gauntlet. Hence, the Soviet Union does not enjoy as much mobility as it would like in the region.

In a tactical sense, the Pacific Fleet's position in the Japan Sea presents some of the same negative features as the Black Sea Fleet and Baltic Sea Fleet face in their waters. All are bottled up by contiguous territory. However, the situation for the Pacific Fleet is qualitatively worse because its regional and global missions are more akin to those of the USSR Northern Fleet. But the Northern Fleet—for all the obstacles it faces in terms of weather and NATO adversaries—still has comparatively free access into open waters. The Soviet Pacific Fleet, on the other hand, is relatively landlocked. Only those vessels regularly based at Petropavlovsk are in ice-free waters warmed by Asia's version of the Gulf Stream, the *Kuroshio* (Black Current). However, their site on the remote, and difficult to supply, Kamchatka Peninsula reduces their ability to protect the centers of Soviet activity in the Far East and makes them far from other Asian populated areas. Although U.S. and Japanese naval and commercial vessels ply the Sea of Japan and the Sea of Okhotsk regularly, in part to confirm them as international waters, many observers have long considered the Sea of Okhotsk a virtual Soviet internal "lake" when Moscow wants it to be. It is largely enclosed by Soviet territory, and few reasons exist to enter these waters other than fishing or testing Soviet responses. Though the Sea of Japan clearly is not dominated by a Soviet hinterland, most of its shoreline being the sovereign territories of Japan and the two Korean states, patrolling and transiting Pacific Fleet vessels are so evident in those waters that many Japanese with a grim sense of humor have taken to calling it the "Soviet Sea."[5]

The degree to which the Soviet Union considers the Sea of Japan to be an area under its influence or control is an important question. If one examines the nature of Czarist-cum-Soviet expansionism in Asia, one can make an excellent case that the Soviet Union is motivated by a paranoid variation of Turner's American frontier thesis, which drives Moscow to secure its "frontier" via the creation of "buffer" states that shield the homeland.[6] Though Moscow failed in its efforts to extend its buffer zone of influence to Korea in the early 1950s, Korea—as will be assessed below—remains an object of Soviet efforts. Creation of a buffer zone around the Sea of Okhotsk clearly motivates Soviet obstinacy about yielding to Japanese claims to the southernmost of the Kurile Island chain because those islands help secure the Okhotsk buffer zone. So far the furthest Moscow has been willing to go was to raise the prospect of possible return of the two smallest, and least important strategically, of these islands (the Habomai group and Shikotan), which would leave the two largest (Kunashiri and Etorofu) in Soviet hands and used by Soviet forces.[7] For Soviet purposes they have achieved a de facto buffer in the Okhotsk region, albeit one that is only grudgingly and implicitly recognized by the USSR's adversaries. Moscow has not got nearly that far in its dealings with the Sea of Japan and contiguous states, but its intentions seem to be functionally the same. The USSR has virtually no chance of guaranteeing its control of the Sea of Japan the way it can contemplate for Okhotsk because in the former case it lacks control of the percentage of surrounding territories it enjoys in the latter case. Nonetheless, Moscow seems intent upon pushing its Japan Sea "frontier" as close to the borders of the three other contiguous states as is feasible, and to minimize—by one means or another—the ability and/or desire of those three states to obstruct Soviet use of the Sea and opportunity to come and go as it pleases.

The latter concern revolves around maintaining free passage through the relatively narrow straits which allow access to the Japan Sea. Though all the states involved tout the virtures of freedom of the seas, they all also are sensitive to those straits' strategic utility as so-called chokepoints. By emphasizing and preparing for that role in a crisis, the United States and Japan tacitly acknowledge the watery frontier and buffer Moscow seeks in the Japan Sea, but far more grudgingly than they do vis-à-vis the Okhotsk Sea. In the Japan Sea, the United States and Japan routinely work to keep the principle of free seas alive and well. South Korea, to a lesser extent, does the same thing and supports the United States and Japan in their principle efforts. However, it is a difficult task and one which must confront the advantages the USSR Pacific Fleet enjoys in its front yard.

Moscow appears to recognize the abilities of its adversaries to constrain its free access to the Japan Sea in a crisis via blockades and mining of the chokepoints. Its response is two-fold: to build capabilities in the Pacific Fleet to crack those barriers, and to undercut the cooperation between the United States and its allies necessary to manage such an operation. While quietly working on the former and hoping that it can prevail in a crisis,[8] the Soviet Union works in more obvious ways to achieve the latter goal. Soviet relations with Japan and Korea are the subject of other chapters in this volume, but—since they are central to the stability of the Japan Sea—their strategic aspects will be addressed here too.

Moscow's diplomatic efforts toward Tokyo were markedly stepped up after General Secretary Gorbachev and Foreign Minister Shevardnadze entered office. The first shoe was dropped during Shevardnadze's January 1986 visit to Japan.[9] The second was dropped at Gorbachev's Vladivostok speech. This latest phase in Soviet-Japan relations follows a low ebb in bilateral ties, marked by heavy-handed Soviet efforts to intimidate Japan strategically by attacking its defense ties with the United States and lure it into closer economic relations using resource inducements with patently obvious strings attached. Moscow experienced little success in those crude endeavors for reasons which will be addressed in the Japan portion of this chapter. It is too early in Moscow's current campaign to be confident that it, too, will fail, but past precedent suggests that the USSR is not likely to achieve a turnaround in its relations with Japan.

Soviet relations with South Korea are much more ephemeral. There are, of course, no formal ties between the two countries, but—nonetheless—the USSR is able to exert influence over the Republic of Kampuchea (ROK). At least through mid-1988 Moscow's most blatant instrument of influence is its ability to jeopardize the success of the Seoul Summer Olympics. South Korean government officials are inordinately anxious about the prospect that the Soviet Union might yet lead a boycott that could disrupt an event which they hope will be symbolic of their acceptance as an internationally important country. As a consequence of this South Korean anxiety, Moscow has tremendous temporary leverage which forms one part of its "Korea card." Of longer-lasting utility for Moscow is its ability to dangle before South Korean eyes the same long-run natural resource enticements that appeal to some Japanese.[10] What Moscow says about economic affairs to Tokyo is heard in Seoul, too, by officials who know they cannot respond now but might be able to someday if an atmosphere of detente resurfaces. Moscow's criticism of U.S.-Japan defense ties, and especially of the prospect of closer U.S.-Japan-ROK strategic ties, seems primarily designed to appeal to North Korea and to Japanese leftists. However, it also finds an audience in South Korea, where many people are ambiguous about any larger regional defense role for Japan. Ironically, Moscow's criticism of such a possibility echoes in South Korea and provides the USSR with another indirect lever over the ROK. On balance, however, Moscow seems to accept the limitations of all its leverage over South Korea, and rests its limited hopes regarding that country on the unlikelihood of the ROK becoming more of an obstacle to Soviet objectives then it already is.

Soviet relations with North Korea are a much larger and more important topic, much of which is not directly concerned with the Japan Sea area but is focused on Sino-Soviet tensions and Pyongyang's role in their interstices. Out of that long and complicated set of interactions North Korea had emerged as neither fish nor fowl on a spectrum of Sino-Soviet identities. With the advent of a pragmatic Dengist reform group in China, and a slicker, younger, and image-conscious Gorbachev reform group in the USSR, North Korea has been faced with new circumstances with which it is not comfortable. Pyongyang's opportunistic pendulum has swung back and forth between Beijing and Moscow several times before and it has now swung once more toward the Gorbachev regime. Pyongyang's reasons will be examined below, but for now this latest shift will be assessed in terms of what it means to the USSR and the Japan Sea.

Thanks to Soviet offers of improved weaponry and technology to North Korea made during the visits of M. Kapitsa and G. Aliyev to Pyongyang in 1984–85,[11] there is an increased likelihood that the Soviet Union might gain expanded access to facilities in North Korea, which might include access to ports on the Japan Sea and the Yellow Sea. Though some analysts have been alarmed by these developments, and they are potentially serious in certain other respects, in terms of naval affairs they need to be kept in perspective. Soviet access to additional ports on the Sea of Japan (in Korea or elsewhere in the USSR) would not necessarily alter Moscow's basic dilemma caused by constraints on Soviet mobility imposed by the chokepoints. Access to Yellow Sea ports would help divert opposing forces' resources from the Japan Sea, but only at the potentially high price of granting Pyongyang added influence over Soviet strategic interests and providing incentives for the PRC to perceive a PRC-ROK chokepoint to limit access to the Yellow Sea. On balance, therefore, one should not be precipitously alarmed by the apparent strategic gains made by the Soviet Union vis-à-vis North Korea.[12]

Except for the USSR's relations with North Korea, in which strategic cooperation plays in increasing role, contemporary Soviet relations with its other Sea of Japan neighbors seem calculated to reduce their perceptions of the Soviet Union as a source of threats to the region. In these terms Moscow seems intent on portraying the Japan Sea as a means for communication—a seaborne bridge—rather than as a barrier dividing the states which encircle it. Because of Japanese and South Korean strategic dependency upon the United States, the putative bridge cannot be realized. Although South Korea's view of the Soviet threat does not coincide with that of the United States, in part because of Soviet behavior in the area, Seoul is aware of the Soviet threat, as will be seen below. It is in Japan where Soviet rhetoric achieves some of its intended impact.

II. Japan

On the surface Japan's interests in pacifism, low defense spending, and frequently utopian theorizing about international relations seem to make that country ripe for Soviet plucking. Actually, though many Japanese are conscientious listeners to what Moscow has to say, and appreciate the more pleasant demeanor of current Soviet officialdom, few important Japanese have been deceived by the softer sell emanating form Moscow. Like many Westerners, the Japanese see clearly the "iron teeth" between the "nice smile." Perhaps more than most Westerners, the Japanese have a well-developed sense of skepticism when dealing with the Soviet Union. This stems from their history of tension, conflict, racism, mistreatment, and ideological differences ranging from the Russo-Japanese War through events before, during, and concluding World War II and into the ups and downs of the Cold War, detente, and such neo-Cold War events as the Soviet ground, naval, and missile force buildup in the Far East, and provocative actions at sea and in the air aimed wholly or partially at Japan. The callous KAL 007 shootdown and ominous Soviet force deployments and exercises on Japanese-claimed islands off the shore of Hokkaido stand out as graphic symbols of Soviet actions, which speak much louder to the Japanese than mellow Soviet rhetoric.[13] Interestingly, Soviet private views of Japan fully reciprocate such suspicions.[14]

Consequently, efforts by Moscow to present a benign face to Japan, focusing on alternative paths to peace and security and developing Japan-Soviet goodwill in cultural and trade relations, which were well exemplified by Gorbachev's September 1985 discussions with visiting Socialist Party Chairman Ishibashi Masashi, have not been very persuasive to many Japanese.[15] Similarly, the Soviet Union has enjoyed only marginal success in its efforts at cultivating the friendship of Hokkaido residents by establishing culturally oriented "friendship centers," inviting delegations of subsidized Japanese visitors across the Japan Sea, and dangling economic and fisheries enticements before Japanese eyes.[16] The Soviets still have a long way to go in persuading this target population why they should alter their fears since Hokkaido is as close to a front-line milieu as Japan possesses. It lies on the fisheries "front," the Northern Territories "front," and—as the presumed pathway for any ground assault from the north—it is what passes for Japan's military "front," too. Residents of Hokkaido traditionally have been receptive to conservative Japanese arguments about the need to stand up to Soviet pressures and intimidation. In part this receptivity stems from some residual anxiety in Hokkaido about its very "Japaneseness." As the last settled portion of Japan, it—along with the Kurile Islands and Sakhalin (which many Japanese still refer to as *Karafuto*)—shares a certain common heritage of insecurity and shallow roots. Soviet possession of Sakhalin and the Kuriles has underscored the unease of Hokkaido's long-term place in the scheme of things Japanese. Japanese treatment of Hokkaido as its last frontier, with exotically un-Japanese qualities, reinforces such anxiety among its residents. To the extent these attitudes bolster their desires to solidify their Japanese identity by distancing themselves from the Soviets and rejecting Moscow's overtures, it is a positive factor. However, Hokkaido's sense of isolation from "mainland" Japan also, and perversely, makes its residents vulnerable to anyone who seems sincere in their attention toward them. In these terms, the Soviets seem to be making some headway.

By and large, however, the Soviet Union is not making much progress in dissuading Japan from its chosen path in league with the West. Although the level of Japan-USSR trade appears ready to experience an upsurge, it is very unlikely to disrupt seriously Japan's pronounced orientation toward the free world economic system.[17] The Soviet Union simply cannot offer enough economic inducements to Japan to warrant a tilt by Tokyo toward Moscow. Moreover, Japan's economic proficiency, while an appealing counterweight to the USSR's torpid economy, is also a powerful competitor with which Moscow's economic planners would have a difficult time competing. Even given the limited bilateral trade in which they now engage, Moscow, in the person of Alexsey Antonov, Deputy Chairman of the USSR Council of Ministers, on a visit to Tokyo, has already complained about Japan's advantages and a trade imbalance.[18] It does not take much imagination for Tokyo to grasp the sorts of pressures Moscow would exert on Japan should the volume of Japan-Soviet trade blossom and displace Japanese trade with the West. Tokyo does not want to gratuitously provide Moscow with such leverage and, hence, responds very cautiously and slowly to Soviet economic overtures.

The situation is similar on the strategic side. Though Japan may not seem as if it is making rapid strides in its rearmament from the perspective of conservative Americans or Japanese,[19] the Japanese regularly are accused by the Soviets of falling under the influence of foreign (i.e., U.S.) and domestic imperialists, who jointly are propagating some sort of contemporary clones of prewar Japanese militarists and are about to loose these monsters on Asia's unsuspecting masses. This patent rubbish is rejected by most Japanese, though it sometimes finds a receptive audience elsewhere in Asia where people harbor similarly anachronistic misperceptions of the Japanese. Despite such factors and transparent Soviet efforts to sow discord between the United States and Japan, Tokyo has no illusions about the source of its security. Consequently, Japan has not permitted Soviet entreaties or threats to alter its fundamental reliance on U.S. commitments. Although some of us in the United States urge an accelerated pace, some progress is being made within Japan's self-imposed constraints. These include moves by Prime Minister Nakasone to rearrange Tokyo's priorities toward an emphasis on air and naval forces (away from ground forces) and to participate in the United States' vaunted SDI research programs.[20] Such efforts fly squarely in the face of Soviet efforts to get Tokyo to shift away from measures that nudge Japan into closer cooperation with the United States as it copes with the Soviet Union.

Japan very clearly recognizes the possibility of a threat emanating from the Soviet Union, but it prefers not to focus on that threat in ways that would feed Soviet propagandists or raise American hopes that Japan is about to pick up more of the common burden. Instead, Tokyo hopes to reinforce its existing reliance on the United States in ways that do not unduly excite either superpower. Consequently, Japan has undertaken an obligation to assume expanded self-defense roles that incorporate the concept of limited sealane defenses. Though this has caused some anxiety in Japanese left-of-center, pacifist, and parsimonious circles, who see it as akin to the proverbial camel's nose under the tent flap,[21] Japan has begun assuming a somewhat larger role. From the perspective of Japan Sea strategic affairs, enhanced Japanese self-defense clearly has major implications for its creation of capabilities to assure control over and—if necessary—closure of its domestic straits and cooperation with the United States and ROK to do the same thing in the Soya, Tsushima, and Korea straits. Japan's assumption of responsibility for, and creation of the wherewithal to accomplish, the defense of certain SLOCs (Sea Lines of Communication) will greatly aid its ability to help cordon off the Japan Sea in a crisis. Such duties are not as clearcut as they probably should be, but the fact that they have been undertaken by Japan demonstrates where Tokyo stands in relation to the Soviet Union and its threat to the Japan Sea.

It is reflective of Japan's reality, and symbolic of its attitudes, that Japan's position on the Japan Sea configures most of its harshest and least developed territory (in Hokkaido and northern Honshu) toward the Soviet Union. In effect, Japan's backside is turned toward the Japan Sea, washed by the cold waters of the *Oyashio* (Kurile Current) out of the Okhotsk. This is the least hospitable part of Japan. In contrast, Japan's outward-looking, highly developed, and densely populated territories are primarily facing the Pacific. This analogy should not be taken too far because it is, of course, important that Japan not let its less developed areas lag behind terminally, but for now the contrasts are telling. The exception that tends to prove that rule is the relative prosperity of the southern portions of Japan's "backside" which face South Korea compared to those which face North Korea and the Soviet Union. Climate certainly plays a major role in such contrasts, but on political and economic terms the comparisons are instructive.

III. South Korea

The proximity of southern Japan to South Korea across the Sea of Japan ("Eastern Sea") clearly has played a role in the ROK's growing prosperity. What is far less understandable is the relationship between South Korean security and Seoul's perceptions of threats emanating from the Japan Sea. South Korea is understandably preoccupied with the threat posed to its security by the

STABILITY AND INSTABILITY IN THE SEA OF JAPAN 83

Kim Il Sung regime in North Korea which—despite periodic "peace campaigns"—has never forsworn its ambition to achieve unilateral unification by force on its terms. Such a palpable threat certainly does tend to focus one's mind on the task at hand. Accordingly, South Korea devotes almost all its defense energies to the North Korean threat—on the ground (and occasionally under it!), in the air, and at sea. While no one could legitimately question such priorities in the past when the ROK was struggling to keep its head above water economically, politically, and strategically, that is no longer so true. The ROK has matured to the point where it could afford to look further afield in security terms. In part such expanded horizons could include cooperation with the United States and Japan in defense of common interests.[22] That is an issue to which we shall return in conclusion. However, before doing so and before considering North Korea's place in this mix, it is worthwhile examining why Seoul does not get more agitated about broader threats to its security and why this is important for Japan Sea stability.

There are two clear examples of such broader threats. One, though interesting, is beyond the scope of this chapter, namely, the growing threat to South Korean security posed by economic factors. Just as economic factors contributed to the emergence of a more stable and economically secure ROK in the 1960s and 1970s, the 1980s have witnessed the growth of resource- and market-oriented threats that could jeopardize the ROK's prosperity. In this way, the ROK's security interests have begun to resemble Japan's broad definitions. However, Seoul's view of security remains much narrower than Tokyo's, producing a false duality in Seoul's policy and contributing to the ROK's sense of insecurity. More relevant here is the nature of Seoul's perception of the threat posed by the Soviet Union.

Obviously Seoul is concerned about Soviet threats as they are perceived in Washington. The ROK needs the United States to remain a resolute superpower, holding down its responsibilities as a superpower in Korea's corner of the world. However, officials in Seoul rarely see all such superpower relations as having direct relevance to their immediate fate in terms that South Korea can remotely influence.[23] To be sure, South Korean defense authorities monitor and worry about the increased presence of the

Soviet Pacific Fleet off Korean shores, particularly in the Japan Sea, the Tsushima and Korea straits, and near Chejudo. They also are well aware of Soviet perceptions that an opposing correlation of forces is in the making, designed to constrain Soviet mobility.[24] But in public, ROK officials have been very cautious about encouraging such observations and speculation. A lot of that caution can be attributed to the concerns and aspirations vis-à-vis the Soviet Union that focus on the 1988 Olympics and on a probably vain hope that Seoul can use improved ROK-USSR ties to weaken North Korea's geopolitical position. That is precisely the sort of thinking in South Korea which gives credence to a "Korea card" for Moscow that was noted previously. Though it is impossible to prove, there seems to be another, and overriding, factor causing such concern and the diffidence that is frequently expressed in South Korean private views of the region and the U.S. role there: they cannot conceive of Washington ever asking the ROK to help the United States or of the ROK ever having enough left over after coping with North Korea that it could lend a meaningful hand to the United States. The accuracy of such assumptions will be addressed in conclusion, but for now the result is a ROK which shies away from seriously considering the Soviet Union an immediate threat to South Korea.

Against these considerations and in light of the relatively underdeveloped and less populated status of much of South Korea's eastern coast north of the Pohong-Ulsan industrial area (which makes that northern area functionally similar to Japan's northwestern "backside"), Seoul does not get unduly agitated by de facto Soviet influence over its Japan Sea "frontier-buffer." Seoul's security concerns in this portion of the Japan Sea are narrow ones, namely, protecting its own version of an offshore buffer so that North Koreans cannot use these regions to launch infiltration missions or a diversionary attack. Were the Soviet Union to assist North Korea in such endeavors, Seoul undoubtedly would become agitated. However, that prospect seems unlikely to South Korean leaders, who do not visualize Pyongyang requiring such assistance, much less Moscow providing it. As a result, the only area of the Japan Sea's security which arouses much concern in Seoul (other than North Korean threats) is from Pohang around to Pusan where Korea faces Japan. This is, after all, the site of one of the two most difficult chokepoints (the other being the Soya Strait). In this region the ROK appears more than content to play a corollary role in the U.S. Seventh Fleet's task of keeping the lid on the USSR's egress points. For the moment this limited agenda for the Japan Sea suffices for South Korea and presumably does not cause much concern in Moscow.

IV. North Korea

Unlike South Korea, which can afford to place the Japan Sea's broader security relatively low on its defense agenda precisely because it is high on the United States' agenda, North Korea is compelled to give it a higher billing. In part this is because the DPRK's navy is a two-part affair. Effectively cut off from the sea far to the south by the presence of a South Korean de facto "island" athwart its border, North Korea is forced to maintain two navies, each of which must cope with a difficult strategic milieu.[25] Its East Sea fleet must operate directly against the background of the variables cited previously for Soviet policy toward North Korea. However, to keep its distance and to adhere to its vaunted *juche* policy, Pyongyang is compelled to deal more directly than is South Korea with superpower activities in the Japan Sea. Consequently, North Korea acts more like an independent variable. In that context, two factors seem to concern the Kim regime most.

The factor Pyongyang spotlights with great regularity is its supposed fear of a U.S.-Japan-ROK cabal focusing on it. It is uncertain how serious Pyongyang is in its heated rhetoric, but—if taken at face value—North Korea seems to be deathly afraid of that alleged threat. However, unless Pyongyang's policymakers and their propagandists are totally paranoid or are utterly misreading the evidence of contemporary and/or planned trilateral U.S.-Japan-ROK cooperation, there is no reason to accept their fears at face value. More likely, Pyongyang says what it does about such trilateralism among its adversaries because it hopes to preempt such an eventuality by stimulating Japanese and South Korean mutual enmity. Rather clearly, Pyongyang's fears—paranoid or not—revolve around the United States' ability to inflict retaliation upon North Korea on South Korea's behalf and to use Japan as a base to support its Korean ally. Disrupting such indirect cooperation, much of which focuses on Japan Sea security issues, thus becomes a high priority for North Korea. Along the same lines, North Korean defenses—primarily against South Korean infiltration, but also against the grossly exaggerated possibility of a U.S.-backed invasion—make the Japan Sea a natural avenue for such an attack launched from back-up bases in Japan and Okinawa. In short, realistic or not, North Korea remains concerned about its superpower adversary using the Japan Sea as a theater against it in ways that South Korea has great difficulty visualizing vis-à-vis its superpower adversary.

The second factor which concerns the Kim regime about the Japan Sea focuses on the Soviet Union's strategic need for alternatives to the confinement of Vladivostok. Pyongyang is an old hand at manipulating its major power backers and balances one off against the other. The conceivable advantages Moscow could gain at the PRC's expense by securing a firmer strategic foothold in North Korea give Pyongyang leverage over both Moscow and Beijing, which it uses gingerly but purposefully. Short of a major change in Sino-Soviet relations, Pyongyang must use a fine vernier scale on its leveraging maneuvers to assure that the process does not get out of control. Pyongyang cannot afford to move too far, too fast, lest it precipitate the sort of major change that would reduce the value of its leverage. Precisely because North Korea borders on both the Soviet Union and the PRC, it probably cannot afford to run the risks of going as far as Vietnam has gone in permitting Moscow access to North Korean facilities. The trick for Pyongyang is how to keep the pendulum swinging to and fro without aggravating the occupants on each end of the spectrum, both of which periodically display some testiness toward North Korea.

A key question about North Korea's security policy (in general and vis-à-vis the Japan Sea) is the issue of who will succeed Kim Il Sung and how he reacts to Moscow. Whether or not his son, Kim Jung Il, persists as his successor, the degree of Soviet influence may be crucial. If Soviet cozying up to Kim Il Sung in the mid-1980s strongly reinforces USSR-DPRK defense ties on the eve of Kim's demise, any elements in the Pyongyang hierachy who are pro-Soviet will be in a strong position to sway post-Kim policies. Should that occur and the Soviets make basing gains, it will be at the expense of China, and will add to the uncertainty surrounding Japan Sea stability.

V. PRC

Even though China is near the Japan Sea, is concerned about that area's security, and is a major power, Beijing lacks much say about the fate of that subregion. The PRC is a mediocre naval and air power, its strength being primarily in numbers of ground forces. Consequently, Beijing is a highly interested observer of the Japan Sea strategic scene, with great concern about the Soviet Pacific Fleet's ability to help close the circle around the PRC by a pincer arrangement reaching out from Vladivostok and Petropavlovsk to Cam Ranh Bay. It also is, as already noted, concerned about Soviet gains in North Korea and the possibility of Moscow's circumventing Japan Sea constraints by acquiring bases on the Yellow Sea.

Because of these concerns and its inability to do much about them in terms of unilaterally countering Soviet actions or expressions of influence, the PRC is limited to two basic options. One unacceptable option is to compensate for Soviet measures by yielding or compromising on outstanding issues in Sino-Soviet relations. Instead of that obvious nonstarter, Beijing chooses to rely on what its quasi-ally, the United States, and the latter's friends and allies can do to counter the Soviet presence. This age-old Chinese tactic of using one "barbarian" against another one works just as well today as it ever did—imperfectly. The risks associated with the tactic focus on the chances that one (or more) useful barbarian might get out of hand. Since long-term PRC goals are far from identical with those of the United States, Beijing can be no more comfortable about the necessity of trusting the United States than the United States is about trusting the PRC. It is the parallelism of certain goals, which overlap in part, that binds us together in the absence of real confidence in each other. However, while some Americans harbor fears of fostering a Frankenstein's monster in Communist China, there is no precisely corresponding fear in the PRC about the United States. Chinese fears of that ilk revolve more around either apprehension that China will be subverted by Western luxuries and vices or that the barbarian's barbarian (i.e., Japan) could revert to its evil previous incarnation. It is a sign of the depth of Chinese concerns about the Soviet Union that Beijing is prepared to run the risks of "subversion" and revived Japanese militarism in order to keep the United States on a parallel track.

VI. United States

It is a commonplace of superpower strategic relations for Western analysts to put much more emphasis on the northeastern Atlantic than on the northwestern Pacific. This is understandable because of longstanding U.S. strategic interests in Western Europe and the threat posed to those interests by the Soviet Union's Northern Fleet. This threat became especially acute under Admiral Gorshkov's reshaping of the Soviet Navy into a "blue water" entity. However, in recent years the rapid Soviet naval buildup in the Pacific has posed a potential threat to U.S. interests in the Northeast and Southeast Asian regions where those interests promise to grow to a stature matching the United States' interests in Europe. Therefore, if the Pacific Fleet were not boxed in, the USSR's Northern and Pacific fleets could be considered roughly comparable in size and function. But the Pacific Fleet remains constrained by the Japan Sea, whereas the Northern Fleet must be faced on the open seas. This is what makes defending against the Soviet Union in the Barents and Norwegian seas so difficult. When coupled with the proximity of the Western European centers of population and economics, and with the Soviet Union's similar centers, it is easy to understand why Western analysts assign the priorities they do.

However, continuing this emphasis is very dependent on the ability of the United States and its allies to keep the Pacific Fleet bottled up when deemed necessary. The U.S. deployment of the Seventh Fleet from Japan, maintaining mobile Marine Corps units in Okinawa, and important Air Force elements ranging from Japan and Korea to Guam and the Philippines, all make this corking of the bottle possible. Should anything occur which disrupts or cancels the containment of Soviet forces by means of a crucial

Asian maritime strategy,[26] the implications for the global balance of power would be just as serious—if not more serious—as loosing Soviet forces in the Atlantic. The main reason for this is that the elements of the potential Asian-Pacific correlation of forces on the United States' side versus the USSR, though favoring the United States and its allies versus the USSR & Co., are far more uneven than they are in the Europe-Atlantic theater. China is useful on land, but not much more. The ASEAN states and Australia/New Zealand would be somewhat more useful in their subregions, but only locally. The same could be said of the ROK, if it were to broaden its strategic horizon. Only one state in East Asia truly has the potential to add to the opposing correlation of forces in ways likely to make strategists in Moscow sit up and take serious notice—Japan. Since Japan clearly is the state in the larger region with the most to lose by any setback to U.S. abilities to contain the USSR Pacific Fleet, and is in the best position to help the United States achieve that mission, it is no surprise that a generally reluctant Tokyo has been particularly forthcoming in regard to seeing Japan Sea chokepoints as essential to Japan's self-defense.

What would the failure to secure those points mean? Perhaps the best way to visualize the impact of that prospect is to recall the buffer-frontier notion. If the Soviet Union is able to expand that "frontier" beyond the Japan Sea, where might it draw the line? There is no clear answer, and that is the problem. Should Moscow's oceanic "frontier" be expanded, it will come up against the United States' version of the transpacific frontier embodied in the idea of forward deployments. Though many Asians wonder about the fairness of U.S. bases acting as magnets for Soviet enmity in an era in which the frontiers do not overlap, the possibilities inherent in tensions over an area in which each superpower asserts a "frontier" that overlaps the other's would be even more dangerous. Making this prospect still more troubling is that coping with a USSR Pacific Fleet freed from the confines of the Japan Sea would require vastly more U.S. and allied defense resources, comparable to those now devoted to Atlantic defenses. It is, therefore, far wiser and cheaper to keep the cork in the bottle. The key question, however, is how to do it.

VII. Conclusion

The country in the Japan Sea picture with the most to gain by changing the existing balance clearly is the Soviet Union. Its options for breaking out of its confines, as outlined in the preceding sections, are not promising. Short of a loss of resolve by the United States and its allies, Moscow stands little chance of loosening the grip around the chokepoints. About the best Moscow can hope for is a shattering of the bonds which bind the United States and its allies. Clearly, Moscow will do everything it can to foment such discord.

The United States has less to fear in this regard than it does from the prospect that growing tensions between the United States and its allies over economic and/or political issues will generate the sorts of discord that Moscow can foment. Frictions over trade fairness, protectionism, progress toward political pluralism, and appropriate and fair levels of defense spending already are sowing dissent in U.S. relationships with Japan, South Korea, the Philippines, China, and—to a lesser extent—virtually all the other states in the Western Pacific. Such problems, if allowed to fester, could easily serve Moscow's purposes. While this should certainly concern United States and allied policymakers, their concern should not lead them to paper over the problems where they might fester less visibly. Instead, it is time that policymakers in all these countries recognize and act upon the inequalities that divide us. If we all work together to achieve fairness and mutual respect, there is virtually no chance that discord can be exacerbated.

How might this nebulous rhetoric be implemented in the Sea of Japan? Although there might well be other viable paths toward improved cooperation designed to assure the strategic cork will stay in place when needed, I suggest that the fairest course of action would involve burden sharing via power sharing. Although I contend this should apply to larger U.S.-Japan and U.S.-ROK ties, in the Japan Sea it would involve a restructuring of the burdens shared in defending the chokepoints versus responsibilities the United States bears elsewhere from which Japan and the ROK benefit. If Japan and the ROK were to bear more explicitly some of the responsibility for standing up to the Soviet

Union, standing ready to seal those points off if necessary, they would be performing duties somewhat comparable to our NATO allies. However, in both Asia and NATO there are inequities in burdens shared versus benefits received. I suggest that the incentive the United States needs to offer these allies to induce them to lift heavier burdens is to encourage them to accept more responsibility by becoming part of the decisionmaking process.[27]

The difficulties that might be encountered in pursuing burden sharing via power sharing are easily imaginable, but they should not be considered insurmountable. With the proper commitment from all parties, they can be overcome and translated into the foundation for an improved strategy that should enable the United States and its allies to preempt the sorts of discord from which the Soviet Union can now hope to take advantage. Though it would be tempting to lay out the specific responsibilities that one could propose for each party to undertake as part of a joint strategy, that temptation should be resisted. It is precisely the tendency of Americans to tell our Japanese partners how burdens should be restructured, what missions they must assume, and how to deal with sensitive technological issues—all in the name of meshing Japan into a U.S.-designed strategic game plan—that discourages the Japanese from participating as a real partner. To reverse this, and encourage the Japanese to become the kind of powerful ally for which they have tremendous potentials, I believe it is necessary to engage Tokyo in a negotiating process that will produce a joint strategy between strong and mutually supportive allies. The same approach can be used regarding South Korea, though its potentials for helping the United States are markedly smaller. However, the last thing that is needed is more premature advice to these allies about specifics before they, and we, have meshed our threat perceptions and devised a shared assessment of what really needs to be done to counter the Soviet Union. One should remember that the Japanese voice and wherewithal in such shared decisionmaking may well contribute to a strategic reconfiguration that could entail different approaches to coping with and, perhaps, deescalating regional tensions.

In this regard it is worth noting that Japan[28] now and in the future may make valuable contributions to joint efforts at deescalation measures in the Japan Sea. Japan's longstanding postwar proclivity toward minimizing armaments cannot continue to be cavalierly written off as hopelessly utopian. The contemporary Japanese temperament on such issues can add a useful dose of prudence to strategic scenarios that threaten to escalate severely. Japanese proposals toward tension and force reduction may often be unrealistic in the short term, but in the future, as their proposals—if made as a responsible partner of the United States—confront the realities of the Soviet Union's armed presence and ominous ambitions, there is an excellent chance that Japanese leaders will quickly become more pragmatic. Such hardnosed pragmatism, which is barely nascent now, could grow speedily and transform Japan's world view into something approximating the world view of the United States if Moscow behaves—as it almost certainly will—in ways calculated to drive Tokyo in that direction. Japan's potentials to seriously rearm and either devise a unilateral strategic posture or become a truly active partner of the United States are tremendous. Moreover, those potentials are clearly recognized by the USSR, which now alternates between entreaties and browbeating the Japanese not to follow such a course.[29] Unfortunately for the USSR, but fortunately for the United States, Japan is edging gradually in that direction. If Japan proceeds far down that road, the nature of the superpower balance in Asia certainly, and perhaps worldwide, will be altered to Moscow's disadvantage.

With such a prospect staring it in the face, Moscow should be able to see the advantages to the USSR of preempting such a budding U.S.-Japan strategic coalition. "Preemption" could, of course, involve a panicky Soviet reaction designed to nip Nippon in the bud before the imperial chrysanthemum flowers again or the "sword" once more is wielded by late twentieth-century samurais.[30] The trouble is that Moscow cannot take any such measures without certain retribution from Japan's American ally. Hence, Moscow's safest course of action by far would be to try to preempt Japan's armed development by proffering meaningful strategic concessions in the region. The emerging U.S.-Japan relationship (or, even more destabilizing, the potentials for a Gaullist Japan to go it alone) should provide ample incentives for the Soviet Union to negotiate an arms control regime in the Japan Sea region and environs. Consequently, although the regional arms race probably will get worse before it gets better, the ingredients are present in this situation to create circumstances conducive to an ultimate reduction of tension if the United States and its allies/friends are skillful in manipulating these delicate relationships.

One major obstacle in the path of tension reduction in the Japan Sea region is the way in which the United States and its allies ultimately decide to treat the strengthened USSR-DPRK military relationship. There are important segments among influential U.S. conservatives who are ready to concede that relationship to Soviet domination. They assume that Moscow will get whatever it wants from North Korea in terms of increased access to bases and transmitting rights, that Pyongyang will become a puppet of Moscow, and that—under Soviet influence—North Korea will become a new outpost of Soviet-style imperialist expansionism.[31] While no one should be naive enough to ignore that possibility, it clearly has not happened yet; nor is it inevitable. If the United States, Japan, and the ROK react to USSR-DPRK ties in a sophisticated manner, there is every likelihood that they can appeal to Pyongyang's palpable unease over too close ties with the Soviet Union. Pyongyang has alternatives to Moscow, does not want to be under its thumb, and should not be written off prematurely. Consequently, the development of USSR-DPRK ties should not panic the United States, Japan, or the ROK into gratuitously assuming the worst. North Korea can still be weaned away from the USSR, probably toward the PRC, and perhaps toward Japan and the West. Tension reduction remains possible vis-à-vis North Korea, if the West does not forfeit the possibility.

Ironically, the prospects for implementing such options based on meshing U.S.-Japan strategic interests may well be dimmed because of reluctance to share power by the country that is likely to be its greatest single beneficiary: the United States. Although each ally's leaders would benefit by the enhanced power they would enjoy, and their country's interests would be better served, the collective reciprocal benefits accruing to the United States would undoubtedly be greater than the separate gains made by its newly strengthened partners. As beneficial as such a policy could be for all parties, tremendous unease over the idea of giving U.S. allies any significant voice in joint policies is likely to impede its acceptance by Americans. Nevertheless, it is a worthwhile goal toward which Americans and Asian allies ought to strive.

Notes

1. *Soviet Military Power 1985* (Washington, D.C.: U.S. Department of Defense, 1985), p. 107.
2. Except where noted, this chapter will use the word "strategic" in its generic and nonnuclear sense.
3. For insightful analyses of these pronouncements, see the articles by Takashi Oka, Roy Kim, Edward Girardet, and Joseph C. Harsch in the *Christian Science Monitor* (cited hereafter as *CSM*), 1 August 1986, pp. 9, 11–12. See also *Far Eastern Economic Review* (cited hereafter as *FEER*), 14 August 1986, pp. 30–40.
4. Those who are unfamiliar with these classics of geopolitical theory would do well to consult the following works for some valuable insights that allegedly outdated theorists can still offer us: Halford J. Mackinder, *Democratic Ideals and Reality* (New York: Holt, 1942); Nicholas J. Spykman, *The Geography of the Peace* (New York: Harcourt, Brace & Co., 1944); and Alfred T. Mahan, *The Problem of Asia and Its Effect Upon International Relations* (Boston: Little, Brown, 1900).
5. *The Asian Wall Street Journal Weekly* (cited hereafter as *AWSJW*), 30 December 1985, p. 17.
6. Malcolm Makintosh, an adviser on Soviet affairs for the British Cabinet Office, offered some insights into this frontier-buffer concept in his 21 March 1985 presentation before the Woodrow Wilson Center's Kennan Institute; see *Soviet Military Presence in East Asia and the Pacific: Implications for Future Western Policy* (Washington, D.C.: The Wilson Center, 1985), pp. 20–21.
7. *FEER*, 9 January 1986, p. 13.
8. An estimate of their success in this effort is meaningless in an unclassified analysis because details of such capabilities entail very closely held information. However, one must note that even those with access to such data can only make an educated guess that would be very subject to the numerous unpredictable variables of warfare.
9. For background coverage of that visit, see *FEER*, 17 October 1985, pp. 30–31 *CSM*, 3 December 1985, p. 24 *AWSJW*, 30 December 1985, p. 7. For an analysis of the visit see *FEER*, 30 January 1986, pp. 26–31.

10. Soviet specialists in Korean affairs were remarkably candid about these potentials during discussions the writer had at the USSR's Oriental Institute and Institute of Far Eastern Studies, 2–3 July 1986, in Moscow.
11. *FEER*, 20 June 1985, pp. 32–34, and 7 November 1985, pp. 18–19; *Sankei Shimbun*, 28 June 1985, p. 1.
12. The writer addressed these issues at greater length in his paper, "The Naval Forces of North Korea, Vietnam, and Cambodia," presented at the Center for Naval Analysis Conference on "The Soviet & Other Communist Navies: The View from the Mid-1980s," 14 November 1985. This also appears as a chapter in James C. George, ed., *1985 Sea Power Forum* (Annapolis, Md.: Naval Institute Press, 1986).
13. For general perspectives on Soviet relations in the region, see (in addition to the Falkenheim paper in this series) the insightful joint U.S.-Soviet study by John J. Stephan and V. P. Chichkanov, *Soviet-American Horizon on the Pacific* (Honolulu: University of Hawaii Press, 1985). For a useful article on overall Soviet intentions in Asia against which its Japan campaign is being played out, see *CSM*, 13 January 1986, pp. 1 and 14.
14. Such views were strongly expressed to the writer during a series of discussions in Moscow with a variety of Asia specialists, 2–8 July 1986.
15. For skeptical press reactions to the Ishibashi talks in Moscow, see the *Asahi Shimbun*'s editorial on 23 September 1985, p. 5, and its coverage of Ishibashi's reactions on 27 September 1985, p. 2.
16. For analyses of this Soviet effort in Hokkaido, see *AWSJW*, 12 August 1985, p. 6, and *CSM*, 16 September 1985, pp. 9–10.
17. For a more pessimistic analysis of what this trade could mean to Japan's Western trade partners, see *FEER*, 26 December 1985, pp. 51–52.
18. *Japan Times*, 15 May 1985, p. 1.
19. See the writers' *U.S.-Japan Strategic Reciprocity* (Stanford: Hoover Institution Press, 1985) and Kataoka Tetsuya's "Japan's Defense Non-Buildup: What Went Wrong?", in *International Journal on World Peace* (April–June 1985), pp. 10–29.
20. *AWSJW*, 23 December 1985, p. 17, and 30 December 1985, p. 17; *FEER*, 25 September 1986, pp. 28–29.
21. A representative example of such concerns is the *Asahi Shimbun*'s editorial in the *Asahi Evening News*, 26 July 1985, p. 5.
22. The writer has addressed this previously in "Nichi-bei-kan sogo anpo taisei o nozomu (Desiring a Japan-U.S.-Korean Mutual Defense

System)," *Chuo Koron* (February 1983); "Security in Northeast Asia: A Trilateral Alternative," *Naval War College Review* (January–February 1985); and "Determinants of Strategic Burdensharing in East Asia; the U.S.-Japan Context," *Naval War College Review* (May–June 1986).

23. For a high-level example of a South Korean perception of Soviet collusion in support of North Korea, see the report on President Chun Doo Whan's interview with *Washington Post* and *Newsweek* journalists in the *Washington Post Weekly*, 10 February 1986, p. 16. While it is important that the leaders of the ROK see a "threat," it is disconcerting that the threat remains cast in the terms of what Soviet dominance in Korea would mean for the United States. Though this approach is, of course, understandable as a lever in ROK-U.S. relations, it seriously underplays the potentials for ROK cooperation with the United States (and Japan) in defense of mutual security interests.
24. For coverage of both Korean concerns, see *The Korea Herald*, 3 December 1985, p. 1, and 1 January 1986, p. 3.
25. See "The Naval Forces of North Korea," *op cit.*
26. For a useful brief overview of contemporary U.S. naval strategy, see James A. Barber, Jr., ed., *The Maritime Strategy*, (Annapolis, Md.: Naval Institute Press, January 1986).
27. The writer develops this concept in global terms in his "Strengthened Western Alliances: Reforming U.S.-Japan Defense Ties," in *The Journal of Contemporary Studies*. (Fall–Winter 1985).
28. South Korea, too, in the future may be able to play a limited role of the sort projected here for Japan. See "Determinants of East Asian Security" for further comments in this regard.
29. See note 14.
30. For those not familiar with the metaphor, it is, of course, drawn from Ruth Benedict's now classic *Chrysanthemum and the Sword* (Boston: Houghton, Mifflin, 1946).
31. For excellent examples of such thinking, see many of the papers presented at the International Security Council's "Conference on the Soviet-North Korean Alliance," Seoul, January 18–20, 1987. See the conference proceedings, but especially the final statement of the conference organizers from which the author, who was a participant, dissented because of its treatment of this issue.

5. The Korean Peninsula Conflict: Equilibrium or Deescalation?

Young Whan Kihl

THE Korean peninsula is a security flashpoint in the Asia-Pacific region. As a heavily armed peninsula, divided Korea continues to act not only as the focal point of armed confrontation between the two hostile regimes and states—the Democratic People's Republic of Korea (DPRK) in the North and the Republic of Korea (ROK) in the South—but also as a strategic fulcrum among the four major world powers maintaining active interests in and surrounding the Korean peninsula, i.e., the United States, the Soviet Union, China, and Japan. The intersection of these two contending forces and prevailing trends, the inter-Korean rivalry, and the major power relations between them create a situation of real and potential regional conflict, making the Korean peninsula one of the most sensitive security barometers in today's world politics.

Any discussion of the future plans for resolving the Korean conflict must therefore take into account the interplay of two basic factors: the evolving strategic environment outside Korea, and the ever-shifting military balance, capabilities, and intentions between the two hostile Korean states. The present chapter will focus on each of the following dimensions of the Korean conflict: (1) the emerging patterns and trends in the strategic environment surrounding the Korean peninsula; (2) a comparison of the military, economic, and political capabilities of North and South Korea; (3) future prospects and speculations regarding the security options open to each state; and (4) certain policy measures considered necessary for deescalation and an eventual resolution of the Korean conflict.

I. Strategic Environment Surrounding the Korean Peninsula

The Korean peninsula (the size of the state of Minnesota) constitutes a zone of conflict in Northeast Asia primarily because of the systemic factor of Korea's geopolitical location. Divided nationhood since 1945, and the emergence of the two separate regimes in contest (across the DMZ that bisects the peninsula into two halves), have also added complexity to the Korean situation.[1] Of the total of 62 million people in 1986, approximately 42 million reside in the South and about 20 million in the North.

Korea's strategic value and importance have been noted historically in the past as a "land bridge" between Japan and China.[2] More recently, the geostrategic value of the Korean peninsula has come to be recognized increasingly as a result of the Cold War rivalry between the United States and the Soviet Union. The U.S. involvement in the Korean War of 1950–53 was the turning point in evolving new U.S. Asia policy to contain the Communist expansionism. East Asia is one of the three "central strategic fronts"—together with Western Europe and the Middle East—according to the former U.S. national security adviser in the Carter administration, Zbigniew Brzezinski, and the Korean peninsula happens to be the strategic nexus in the superpower competition in East Asia.[3] Korea shares with Japan an important waterway called the Korean Strait, or Tsushima Strait, which is perceived by the United States as one of the sixteen important chokepoints or sealanes to contain the Soviet global strategic expansion.[4]

Korea's security environment is thus influenced by the interacting policies of the four major powers with interests in regional stability on Korea. All of these powers publicly support the reunification of Korea as a long-term goal, but none of them would wish to be involved in a future armed clash or war to end the present division of Korea.[5] The four major powers may thus be said to pursue—as a minimum—the common policy objective of maintaining the status quo and regional stability, thereby preventing the recurrence of armed hostilities on the peninsula.[6] Within this broad policy consensus, however, the major powers have attempted separately to render active military support to their respective Korean allies, so as to improve their security and diplomatic status vis-à-vis the opponent.

The pivotal role of the Korean peninsula in global politics, in terms of the cross-cutting of competing interests shared by the four major powers in the region, may be appreciated in the following description by an Australian scholar:

> Northeast Asia is an area of dangers to world peace because it provides the nexus between four great powers with competing ambitions: the Soviet Union, determined to develop the resources of Siberia and to have unimpeded access to the Pacific for mercantile shipping and the projection of naval power; China, determined to be influential over its continental sphere; Japan, a maritime power, lying across the Soviet exits and dependent on the US for protection against Soviet hegemony; and the US, dependent on Japan for its Western Pacific strategic presence. *The Korean peninsula lies at the nexus, manifesting by its division the competing ambitions, pulled and pressed within and without, a self-propelled pawn in a complex power game.*[7]

This vivid and poignant depiction of Korea's geopolitical predicament, ventured in the 1970s, seems equally apt and valid in the 1980s and for the foreseeable future.

Recent Trends in the Major Power Relations vis-à-vis Korea

In assessing the emerging patterns and trends in the strategic environment of Northeast Asia, it is necessary to isolate the key variables in the major power equations surrounding the Korean peninsula. Five factors are especially noteworthy, as a 1983 Rand Corporation study argues, as the likely sources of influence on the evolving strategic environment in East Asia, particularly regarding future U.S.-Korean security relations in the remainder of the 1980s.[8] These are: (1) the great power military balance, especially the Soviet military buildup in East Asia, and Soviet policies toward the Korean peninsula; (2) the vagaries of the Sino-Soviet split and U.S.-USSR-PRC triangular relations; (3) the character of the Japanese-American relationship and the nature of the role of Japan; (4) the evolving political, economic, and military situations in both North and South Korea; and (5) the probable role of the United States in the region.[9]

Regarding the security environment in the remainder of the 1980s, several principal conclusions may be drawn from the preceding analysis: (1) the heightened geostrategic importance of East Asia to the Soviet Union and its continuing military buildup creating the need for compensatory actions by the United States and its allies; (2) the continued reliance on a "swing strategy" by the United States for guaranteeing regional security, which is becoming increasingly risky; (3) the likely evolution of Sino-Soviet relations limiting the ability of the United States to interest China in significantly expanded security cooperation and making the task of maintaining security in the region more complex; and (4) recent trends in several important areas in North and South Korea increasing the possibility for destabilizing developments on the Korean peninsula in the remainder of the 1980s.[10]

Developments in 1984–86 attest to the important role the Soviet Union plays in the region's stability, including the future of the Korean peninsula. The Soviet Union and North Korea are cementing their alliance relations as a result of the Soviet Union's decision to provide North Korea with the latest advanced weapons, in exchange for North Korea's granting over-flight and port-visitation privileges to the Soviet Union. Sino-Soviet relations are likely to enter a period of accelerated "thaw" as a result of the Soviet Union taking new initiatives toward China, as revealed by Soviet leader Mikhail Gorbachev's July 1986 Vladivostok speech and by Chinese leader Deng Xiaoping's favorable subsequent response.

The change in the external environment provides the context within which the two Korean regimes must formulate their respective policies, both domestic and foreign. It is therefore important to know the perception shared by the two Koreas' ruling elites regarding the nature of the strategic environment impinging upon their respective national interests. From the perspective of 1985–86, the emerging trends in the strategic environment in East Asia which bear upon the Korean peninsula seem to favor South Korea over North Korea. The northern triangle involving USSR, China, and North Korea, in spite of the recent signs of Sino-Soviet rapprochement, seems to be less stable than the southern triangle involving the United States, Japan, and South Korea.

Four recent trends involving North Korea, as listed below, are matters of particular concern to the DPRK leadership. These trends are generally perceived by the interested parties in Korea to be less favorable to Pyongyang than to Seoul:

- North Korea's increased dependence on the Soviet Union, for both military hardware and economic trade.[11]
- DPRK-PRC policy consultation decreasing in intensity and frequency.
- North Korea's failure to normalize relations with the United States, due to its continued bellicose anti-American stance.
- DPRK-Japanese official ties failing to materialize, in spite of its desire to increase trade relations.

In October 1986, President Kim Il Sung flew to Moscow to consult with Soviet leader Gorbachev, Kim's second Soviet visit in just over two years. No joint communiqué was issued at the end of Kim's visit, but diplomatic sources speculated that North Korea was clearly anxious to reaffirm Soviet support of Pyongyang's stand on several pressing issues, including reunification, the 1988 Seoul Olympics boycott, inter-Korean dialogue, and North–South Korean military balance. Prior to Kim's departure for Moscow, Chinese President Li Xiannian paid an official visit to North Korea, but his trip was largely ceremonial, producing no tangible results to indicate improved bilateral relations between the two countries.

Four recent trends involving South Korea, on the other hand, are likewise matters of some concern to the ROK leadership. These trends are generally perceived by the interested parties to be more favorable to Seoul than to Pyongyang:

- South Korea's alliance relations with the United States strengthening, in spite of the growing trade friction.
- ROK-Japan economic and diplomatic ties growing, in scope and intensity.
- ROK-PRC trade and informal ties (e.g., sports) expanding in scope.
- ROK-Soviet informal cross-contacts increasing in frequency.

Late in September 1986, Japanese Prime Minister Nakasone visited South Korea, a visit timed to coincide with the opening of the Asian Games in Seoul. China sent one of the largest delegations to take part in the Asian Games, transporting its team members on a direct flight from Beijing to Seoul. The Japanese leader reportedly took the position of representing Seoul's interests and stand on easing tensions on the Korean peninsula during his subsequent meetings with China's Deng Xiaoping in Beijing in early November 1986. Deng evidently told Nakasone of his satisfaction about the friendly reception which Korean spectators gave to Chinese athletes during the Asian Games in Seoul. Regarding ROK-USSR relations, some of the South Korean cargoes destined for Europe are reportedly handled via the trans-Siberian rail- and airways, and Soviet–South Korean trade and joint ventures are openly discussed as a possibility. There is discussion that the Soviet Union desires to use South Korean port facilities for repairing its ships in distress on the high seas.

These recent emerging trends and developments in the external environment of Korea will, over time, influence and color the decisionmakers' perception of the changing reality, thereby dictating the two states' respective security policies in the future.

Recent Developments in North-South Korean Relations

The future resumption of the North and South Korean dialogue will affect both Seoul and Pyongyang equally in the days ahead. It will be perceived, however, to be slightly more favorable to Seoul than to Pyongyang in the long run.

The Korean peninsula zone of conflict, often depicted by observers as a frozen glacier, witnessed some movement toward thaw and lessening of tensions in 1984–85. To "move the glacier," both states took important initiatives in 1984 to resume the dialogue (suspended since 1973), with tacit support by their respective allies.[12] While Seoul responded positively to the North Korean gesture of goodwill in September 1984, in connection with an offer to deliver relief goods to flood victims in the South, Pyongyang adopted a posture of new pragmatism on economic issues, including opening its door to the outside world.[13]

The inter-Korean dialogue, however, was suspended unilaterally by North Korea in January 1986. Pyongyang announced that it would boycott all future sessions unless the scheduled U.S.-ROK annual military exercises known as "Team Spirit" were suspended immediately. Then in July 1986 North Korea proposed a high-level three-way meeting of military commanders in Panmunjom, involving South Korea and the United States, to eliminate the danger of war and to reduce tensions on the Korean peninsula. An analysis of the latest rounds of the dialogue, brief in duration but intense and wide-ranging in scope, reveals underlying motivations and calculations on the part of the respective regimes.

The North-South Korean dialogue and negotiations in 1985 were conducted at four separate levels: Red Cross talks, economic talks, parliamentarian talks, and sports talks. In addition, there were rumors of an exchange of the secret envoys between Seoul and Pyongyang in 1985, to arrange for a possible summit between North Korea's President Kim Il Sung and South Korea's President Chun Doo Hwan.[14]

The Red Cross talks were held to help reunite an estimated 10 million dispersed families scattered throughout North and South Korea. These efforts culminated in the historical exchange of three-day mutual visits by delegates of fifty families from each side on September 20–23, 1985. The Red Cross exchange was also accompanied by the mutual visits of artist troupes from both sides.

The economic talks were held in Panmunjom four times between November 1984 and October 1985, to consider possible barter trade and mutual resource development between the two sides. These meetings, however, were disappointing from the standpoint of producing any concrete and tangible results. A series of preliminary meetings of the interparliamentary delegates also was held in Panmunjom to agree on the possible arrangement for a meeting between the two legislative bodies of North and South Korea, so as to adopt the statements regarding the future of Korea.

The sports talks were held under the auspices of the International Olympics Committee in Lausanne, Switzerland, to negotiate the terms of North Korea's possible participation in the 1988 Seoul Summer Olympics. North Korea earlier announced that it would boycott the Olympics, thereby threatening to increase the tension on the Korean peninsula and pressuring other Communist countries likewise to boycott, unless its own proposal for co-hosting the Olympics was accepted. The rumors of mutual visits by the secret envoys, reported by the Japanese and other foreign media, have been officially denied by the spokesmen of both North and South Korea.[15]

It is important to examine what the sources and motives of these brief yet extraordinary changes were in terms of the two states' respective policies and attitudes, and also what the implications of these changes are for the future security requirements of Korea. North Korea's initiative to resume the dialogue in October 1984 was a brilliant move on the part of Pyongyang to sway the deep suspicion of South Korea toward North Korea, following the Rangoon episode one year earlier which killed seventeen high-ranking officials of South Korea's visiting delegation and narrowly missed President Chun Doo Hwan himself. The move was North Korea's attempt to ease over the negative image of itself as sponsoring state terrorism, an image occasioned by the Rangoon bombing in which North Korea was implicated. The Burmese government severed diplomatic relations with North Korea over the incident, followed by the court trial and conviction of the two North Korean officers captured after the bomb blast on October 9, 1983.

The episode of this brief "thaw" in relations in 1984–85, accompanied by North Korea's posture of new pragmatism and open-door policy in 1985, seemed promising for the prospect of further reduction of tension on the Korean peninsula.[16] The situation in Korea was, however, "still a lot like living in a room soaked full of gasoline," as a Western correspondent reported from Seoul.[17] Stability and peace on the Korean peninsula, in short, are partly a function of the major power balance and the external support which the two states are receiving from their respective allies, both militarily and diplomatically.

II. Comparison of the Military, Economic, and Political Capabilities of North and South Korea

Korea's regional stability depends, as a general rule, on the power balance that prevails among the major powers surrounding the Korean peninsula. It also depends, at this point in Korean history, on the military deterrence existing between North and South Korea in terms of their defense preparedness. This section will present comparative assessments of the military capabilities, strength, and postures of each state, followed by assessments of their economic potential and political institutions, to the extent that these sustain and reinforce the military capabilities.

Military Strength

North Korea's military forces in 1986, as estimated by the London-based IISS (International Institute for Strategic Studies), consisted of some 840,000 personnel, with an army organized into 24 infantry divisions, 2 armored divisions and 5 mechanized divisions, including 3,275 tanks; with a navy of 25 submarines, 2 frigates, and 30 high-speed missile launchers; and an air force of some 854 combat planes including MiG-21 and MiG-23 fighter planes. South Korea's military forces in 1986, by contrast, consisted of a total of 601,000 personnel, with an army organized into 19 infantry divisions and 2 mechanized divisions, including 1,300 tanks; with a navy of 9 destroyers, 6 frigates, and 11 missile-carrying launchers, and 2 marine divisions; and with an air force of some 462 combat planes, including F-5A/B/E/F and F-4D/E fighter planes (F-16s were due to be added in late 1986).[18]

North Korea and South Korea typically spend a large amount of money on the military and the arms race. Some 15 to 20 percent of the GNP, on average, is believed to be spent for defense by North Korea; 6 to 7 percent of the GNP goes to defense in South Korea.[19] Up until 1974, North Korea's military expenditures were much higher than South Korea's. Since around 1975, however, South Korea's military spending has outstripped that of the North, although the estimate for North Korea was subsequently upgraded in 1978, based on new U.S. intelligence reports. In 1979, U.S. intelligence data showed, for instance, a rather rapid increase in North Korean military strength since 1971, indicating that the North might have spent as much as 15 percent of its GNP on military expenditures during these years.[20]

This pattern of military spending in North and South Korea reflects the changes in defense policy orientation and security posture of the two regimes. North Korea has continued its policy of military buildup since 1962, under the so-called Four Great Military Policy Lines adopted during the Fourth Korean Workers'

Party Congress in 1962, which contained the slogans, "Arm the entire population," "Fortify the entire country," "Cadetify all the units," and "Modernize the entire army."[21] North Korea in 1986 was said to have deployed 480,000 troops, or almost 60 percent of its 840,000 armed forces, in the forward position near the DMZ, thereby reducing the lead time for a surprise attack to less than twelve hours.[22]

South Korea has enhanced its defense capability and preparedness since 1971 by implementing the Five-Year Force Modernization Plan (1971-76) and the Force Improvement Plans I (1976-81) and II (1981-86). While North Korea under Kim Il Sung sought military superiority over the South throughout the 1960s and 1970s, South Korea under Park Chung Hee's rule worked to achieve economic supremacy over the North during the same period. Since the 1970s, the South has accelerated its efforts on military buildup, so that the military balance between the two states has now been restored in the 1980s.[23]

Economic Potential

The wealth of the two sides of divided Korea, in terms of the per capita distribution of the economy, was quite evenly matched until the early 1970s, when the South started to outdistance the North. The aggregate GNP of South Korea in 1974, for instance, was approximately twice that of North Korea, which meant that the GNP per capita was almost the same between the two societies in the same year. (The population ratio between South and North Korea is approximately 2 to 1). Throughout the 1970s, however, South Korea's economy started to grow at a much faster rate than North Korea's, so that the GNP ratio for the two Koreas in 1980 was estimated to be almost 3 to 1 and that in 1985 almost 5 to 1 in favor of South Korea.[24] Whereas the GNP per capita in the North grew at an average annual rate of 5.2 percent between 1960 and 1976, and a much slower rate thereafter up until 1985, South Korea averaged a rate of 7.3 percent during the same period between 1960 and 1976, and 7.6 percent between 1961 and 1984. In South Korea, the GNP per capita increased from U.S. $590 in 1975 to $810 in 1977, $1,500 in 1981, and just over $2,000 in 1985; in North Korea, it changed from U.S. $620 in 1975 to $700 in 1977, $950 in 1981, and just over $1,000 in 1985.[25]

The economies of both North and South Korea underwent major structural changes in the 1970s. Both Koreas, for instance, advanced from largely agricultural economies in the 1960s to semi-industrial economies in the 1970s. In the South the share of agriculture declined from 40 percent of the GNP to 20 percent between 1965 and 1976, while industry's share increased from 16 percent to 36 percent.[26] Although the details are unknown, trends in the North are believed to have followed a similar course.

Since the differential pattern and rates of economic growth have far-reaching implications for the future of inter-Korean relations, the probable causes for such consequences need to be identified. Many reasons are given as to why the GNP grew much faster in South Korea than in North Korea in the 1970s. According to a U.S. government study, three factors were responsible for the South outperforming the North in the decade prior to 1976.[27] First, the South spent proportionately much less on defense than the North; second, the South, by importing more efficient technology, had a much higher rate of return on industrial investment; and third, the South developed a dynamic, export-oriented economy that generated the foreign exchange necessary to finance rising levels of capital imports.[28]

The study also noted the structural contrast between the two Koreas. North Korea is "a tightly closed society with a planned economy with many elements of the bureaucratic Soviet model of the 1940s and 1950s," and "its educational system spends about as much time imparting the ideology of Kim Il Sung as instilling more practical knowledge."[29] The technical competence of North Korea's labor force and bureaucracy suffers as a result, and it remains inferior to that in the South. The economic planners and top businessmen in South Korea are not only well educated, many with advanced degrees from foreign universities, but are providing extensive training facilities for upgrading the technical skills of a diligent labor force. In the early 1970s, firms with more than 200 employees in South Korea were required to provide training for 15 percent of their employees.[30] In the communication field, especially, the South is way ahead of the North. The number of vehicles, radio stations, and television stations, for example, is much greater in South Korea.

Military Postures

What additional considerations, besides the military strength and economic capabilities, underlie the strategic planning and defense postures of North and South Korea? The military balance between North and South is influenced by geography and terrain, apart from the number of men in arms and the weapons count. The location of the respective capitals, for instance, makes an important difference in terms of strategic vulnerability and force deployments. Whereas Seoul, South Korea's capital city, is only 50 kilometers (31 miles) from the DMZ, North Korea's capital city, Pyongyang, is 145 kilometers (90 miles) from the DMZ. With 9.4 million people (1983 estimate), or close to 20 percent of the total population concentrated in it, Seoul is highly vulnerable to a possible surprise attack by the North, and the defense of the capital city is the foremost concern expressed by South Korean government leaders.[31]

For these and other reasons, North Korea would have a significant advantage in the initial days of fighting, provided it achieved tactical and strategic surprise. North Korea has the capability to produce its own weapons, such as tanks and artillery, but it has to rely on outside supplies for strategic items such as fuel. Depending on the duration and intensity of warfare, therefore, North Korea is said to require the storage of enough supplies to continue fighting for approximately thirty to ninety days without being resupplied by the Soviet Union and China. The U.S. command in Korea insists that the military posture of the North is offensive, and cites the forward deployment of troops and the discovery of three "invasion" tunnels as evidence for this offensive war preparedness.[32]

South Korea's military posture is described by Washington as defensive, with U.S. troops playing the pivotal role of deterrence or "tripwire" for possible North Korean attack.[33] Tactical nuclear weapons deployed by U.S. troops in South Korea provide a bulwark against potential aggressive moves by the North, although the danger of the United States becoming a hostage in an armed conflict has led some critics of U.S. policy to urge the removal of nuclear weapons and eventual U.S. troop withdrawal from Korea.[34] Until South Korea achieves self-reliance in defense and full control in command, however, the current defensive force deployment and forward strategy under U.S. supervision is unlikely to change. In view of South Korea's rapid progress in building its own defense industry, and its desire to obtain a nuclear capability of its own, the future military balance between North and South Korea may shift to favor the South, largely owing to the superior performance of the South Korean economy.[35]

North Korea's military strategy in the 1980s is noted for three characteristic features, according to a South Korean national security specialist. These are, first, a "combined strategy of regular and irregular wars," i.e., a Soviet-style military operation and a Mao Zedong-style guerrilla warfare; second, a "strategy of preemptive massive surprise attacks," especially against the capital city of Seoul in a so-called three-day war, blitzkrieg-type operation; and third, a "strategy of quick war and quick decision" by concentrating on swift initial military victory and subsequently waging propaganda campaigns to hold the territory through negotiation.[36] The same expert believes that North Korea's strategy would be to try to avoid direct confrontation with U.S. ground forces, by bypassing its position as much as possible, so as to take civilians hostage. Under such circumstances the U.S. forces could not use their nuclear weapons against North Korean invading forces.[37] He describes this possible scenario of the North Korean attack as follows:

> The North Korean 8th Special Army Group would make a surprise landing of AN-2 light aircraft and gliders at a point south of Seoul to create a bridgehead. An amphibious mechanized unit would come from the west coast and land on the banks of the Han River. From the midwest in front of the DMZ, light infantry would be sent through tunnels to emerge behind the front line and create confusion. Then, tanks and mechanized units would pour in from three sides to either capture or isolate Seoul. Under these new operational arts, the North Korean forces would not have to engage directly with U.S. ground force troops on the central front. In addition to this advantage, the North Korean forces would be able to take ROK civilians hostage, making it impossible for U.S. forces to use their sophisticated weaponry (probably including tactical nuclear weapons) and thus facilitating a political settlement.[38]

Ten North Korean divisions and the 8th Special Corps, consisting of some 300,000 soldiers, are reportedly deployed along the front-line areas near the DMZ, ready to pounce on South Korea, according to a *Japan Military Review* article. Each of the North Korean divisions is said to command an infantry regiment, an artillery battalion, a mortar regiment, a tank battalion, an antitank battalion, and an antiaircraft regiment, while the 8th Special Corps has under its command four reconnaissance brigades, eight light infantry brigades, twenty-three special brigades, three amphibious brigades, and five airborne battalions.[39]

U.S. strategy in the ROK currently calls for strike against the North in the event of a blitzkrieg-type attack on the South. This strategy serves as a potent deterrence to North Korea's possible action against the South, given U.S. air superiority in the region.

Political Institutions

The outcome of the current military and economic competition between the two Korean states will be determined, in the final analysis, by the quality of political leadership and institutions. The respective leadership is expected to translate the economic resources and military capabilities into workable political capital and assets. How to manage the "crossover" in power relations between North and South Korea, emanating from a major shift in economic and military power relations, remains one of the most important policy issues.[40]

A more self-sufficient ROK army, for instance, when the current modernization efforts continue, could have the unintended side effect of decoupling U.S. forces from South Korea, thereby destabilizing the region more broadly.

A 1985 Rand Corporation study of the comparative "capabilities" of North and South Korea, in terms of their long-term security implications, reveals the following balance sheet:[41]

The South. The principal strengths of the Republic of Korea lie in (a) its abundant and well-trained human resources, (b) its proven economic record, (c) its rising international prestige, and (d) its fear of and defense against a North Korean attack. The principal vulnerabilities of the South, on the other hand, lie in two areas: (1) the fragile state of its political institutions in a dynamic economic and social environment, and (2) its dependence upon external factors, both economically and militarily.[42]

The 1988 Summer Olympics awarded to Seoul is clearly a recognition of South Korea's new international prestige and status.

The South Korean economy has registered impressive average annual growth rates of 10 percent between 1981 and 1986. Although President Chun Doo Hwan's authoritarian rule has encountered growing domestic dissension and opposition, he has announced that he will step down from office in February 1988. If accomplished, this will enable the first peaceful transition of power in South Korea's republican history.

The North. The principal strengths of the Democratic People's Republic of Korea rest on (a) its tight and absolutely controlled political structure, (b) its potent military establishment, and (c) the absolute control by Kim Il Sung of the economic and social structure. The vulnerabilities of the North, on the other hand, are four-fold: (1) its economic weakness relative to the South, (2) its potential for political instability during succession, (3) its declining international position relative to the South, and (4) the limited support received from its allies, the Soviet Union and the PRC.[43]

Since the Rand study was completed, North Korea has started to receive a greater degree of military support from the Soviet Union in the form of advanced military hardware, including MiG-23s, SAMs, and tanks. North Korean Premier Kang Song San visited Moscow in December 1985 to sign an economic cooperation agreement with the Soviet Union, while the Soviet Foreign Minister Eduard Shevardnadze stopped over in North Korea on his return trip from a visit to Japan in January 1986. These recent developments attest to the growing military and economic cooperation between North Korea and the Soviet Union under its new leader, Mikhail Gorbachev.

The episode of false reporting on Kim Il Sung's death in November 1986 indicates how tenuous the situation on the Korean peninsula is and also what could happen in North Korea regarding a power struggle in the post Kim Il Sung era. Although an overreaction by the Seoul authorities to the initial reports was unfortunate, the episode vividly demonstrated the approaching end of an era in Pyongyang and the possible outbreak of serious political upheaval in North Korea, which could ignite hostilities between the North and the South.

North and South Korea as separate entities are relatively small compared with the giant neighboring East Asian countries of China, the Soviet Union, and Japan. Nevertheless, the combined population of the two Koreas—an estimated 62 million in 1986—is sizable enough that a future unified Korean nation would be one of the larger countries in the world, ranking among the first twenty. The combined economies of North and South Korea, with an estimated gross national production of over U.S. $120 billion in 1986, would also make a unified Korea one of the world's economically strong and viable developing countries. Although Korea is relatively small in territorial size and poor in natural resources, the Korean people are diligent and industrious. The Confucian cultural legacy that makes a virtue out of learning has enabled the Koreans to stress scholarship and to achieve a high educational standard for the population at large. With their well-trained human resources, Koreans are capable of building a prosperous modern nation a reasonably high standard of living for the people as a whole. This possibility of a prosperous and unified future Korea will remain a pipe dream, however, so long as the present territorial division and political ideological bifurcation of Korea remains as it is.

III. Future Scenarios and Policy Options

What is the probable future of the Korean peninsula as a zone of conflict? What future scenarios are possible, and what policy options and alternatives are open to the policymakers, in coping with probable conflict situations on the peninsula? A scenario, by definition, is a set of situations that is hypothetical and contrived, but the possibility of becoming a reality may not be ruled out completely.

Four possible future scenarios pertaining to the inter-Korean conflict may be identified:

Scenario 1. Escalation of Inter-Korean conflict

- Further intensification of the arms race, with the possible acquisition of nuclear war capabilities by either or both Koreas;
- Increased military buildup and defense spending, the two

Koreas serving as the major arms exporters to third countries.

South Korea already possesses nuclear reactors and nuclear power-generating capabilities, and North Korea reportedly signed an agreement with the Soviet Union to receive technology for nuclear power generation. While South Korea's defense industries are rapidly maturing, with the joint venture and co-production arrangements with U.S. firms, North Korea's arms and equipment are exported to radical Third World countries (such as Iran) in the Middle East and Africa.

Scenario 2. Maintaining the Status Quo

- Continued confrontation between the two sides, by freezing the current level of armaments;
- Maintaining the current tension levels, but continuously seeking to manipulate tensions for political-security purposes.

This second scenario best approximates the current situation of inter-Korean confrontation in 1986–87. Dialogue and negotiation were suspended unilaterally by North Korea in January 1986, as already noted. In July 1986, North Korea's defense minister proposed to hold a joint military talk in Panmunjom to bring about possible disarmament, without indicating the resumption of multiple channels of dialogue.

Scenario 3. Reducing Tension

- Unilateral first step toward GRIT (Gradual Reduction in Tension) by either side;
- Proposing a series of new measures, including:
    (i) step by step mutual troop reduction,
    (ii) arms control and disarmament,
    (iii) disarming the DMZ Joint Security Zone.

This third scenario, although daring and innovative, remains as a theoretical possibility only. Neither side is likely, at the present time, to take the necessary bold initiative to break the stalemate and defuse the tension on the Korean peninsula, notwithstanding official rhetoric to the contrary.*

Scenario 4. Institutionalizing the Peace Process
by negotiating the adoption of:

- Nonaggression pact between North and South Korea;
- Peace treaty to replace the Armistice Agreement signed following the Korean War (1950–53);
- Promoting economic, social, cultural, sports, and scholarly exchanges, as well as mutual visits, mails, newspapers, etc., between the two sides.

This fourth scenario, although complicated by obstacles at the present time, seems the most "rational and sensible" alternative for both regimes to pursue. After a prolonged process of negotiation, both states must accept the situation of institutionalizing peace on the Korean peninsula as prelude to the ultimate achievement of reunification of their divided land.

All of the preceding situations in these scenarios represent only theoretical possibilities, to be constrained obviously by factors specific to time and place, allowing for the perspectives of the 1980s and Korea as a divided nation-state. The probability of moving toward any one of these scenarios varies, and the probability of witnessing the third and fourth scenarios in particular at this time is not high. Such probability cannot be ruled out completely, however, because there is always a possibility of either Korea's future leadership taking a bold initiative to move the glacier closer to a warmer climate, thereby bringing about a thaw in frigid relations. What seems to be lacking is sufficient political will and determination on the part of a leadership that is committed to the policy of transcending the current arms race and stalemate.

---

*The possibility of a "thaw" on the Korean peninsula reappeared, however, in March 1987 when the U.S. Department of State announced a new policy of permitting U.S. diplomats to talk with North Korean counterparts, if approached by them, at social gatherings. The Seoul government also announced that it would seriously consider North Korea's July 1986 proposal for holding high-level political and military talks.

Strictly speaking, there are two extreme possibilities of "war or peace" on the Korean peninsula. Perhaps the third possibility of "non-war," or the current "stalemate," may also be added as the midpoint between the two extreme alternatives.

Given the emerging trends in relations that favor South Korea over North Korea in terms of the military-economic balance and capabilities examined earlier, the policy options open to North Korea are rather severely constrained. Under the circumstances, the following courses of action may be considered as "rational" policy behavior for the respective regimes.

Both regimes are under the pressure of pursuing a "deterrence plus detente" policy. This will mean that both Koreas will maintain their military posture at the current level, or even at a slightly higher level, but that they will also continue to manipulate the diplomatic symbol of detente with the other side. More specifically, this will mean that North Korea will continue to rely on its strategy of maintaining a forward deployment posture near the DMZ, while seeking Soviet support in the form of the latest weapons and arms delivery. This will also mean that South Korea will continue to strengthen its defensive military posture, possibly by adopting a Force Improvement Plan III, and maintaining an aggressive stance on negotiation. Also South Korea will continue its reliance on allied military support from the United States and on economic and diplomatic support from the United States and Japan.

A recent study of the military competition between North and South Korea speculates about three possible choices confronting South Korea in the immediate future. They are, first, a continuation of the present policy, with 6 percent of GNP defense expenditures; second, an increased defense budget and manpower ceiling; and third, a major shift to reliance on high-technology weapons, with a 25–40 percent manpower reduction.[44] None of these scenarios, according to Sneider, is likely to have an adverse economic or sociopolitical effect on South Korea. The third choice seems the more realistic in the light of the ROK's recent decision to join the United States on the SDI plan.

But North Korea will be adversely influenced by an increase in the South Korean defense effort. Although continuation of the current defense policies in the South (Scenario 1) is not likely to force a change in North Korea's policy, either an increase in the defense effort (Scenario 2) or a shift to reliance on high-technology weaponry (Scenario 3) will exacerbate the dilemma facing the North regarding resource allocation decisions between defense and the economy.

Confronted with these unpleasant situations, North Korea will have basically two options: either a reconsideration of its highly belligerent military posture, or consideration of a preemptive attack on the South.[45] The former seemed to be the more rational option open to the North in 1985–86. From the South Korean perspective, however, the latter option by the North, and therefore the contingency of the worst case of a North Korean-initiated attack against the South, cannot completely be ruled out in 1987, or in the years beyond.

Although most of the preceding scenarios are credible and realistic, the continued jockeying for power and supremacy by either regime vis-à-vis its opponent will guarantee no breakthrough in the current stalemate of the inter-Korean arms race. Therefore, we must consider some new policy measures, which are risky but worth contemplating for the sake of a higher purpose of institutionalizing the peace process on the Korean peninsula.

IV. Policy Measures for Deescalation and Conflict Resolution

The preceding analysis indicates that two basic factors pose obstacles in settling or resolving conflict in the Korean peninsula zone. These are, first, the systemic variable of Korea's geopolitical location, whereby the major powers' interests compete and crisscross; and second, the continued status of Korea as a divided nation, whereby the two hostile regimes are locked into a zero-sum game situation. Nevertheless, it seems clear that the current escalating arms race is not only wasteful but also dangerous to the peace and stability in the region. If we agree on this assessment of the Korean stalemate, we must seek alternative measures for reducing the level of tension and institutionalizing the peace process.

What are the specific measures, if any, for bringing about tension reduction and conflict settlement on the Korean peninsula? Some of these measures, to be discussed at the four separate levels, may be grouped under military, dialogue, diplomatic, and strategic considerations. Both the rationale and the logic behind these measures are self-evident in the light of the arguments already presented, so an explanation and justification will not be repeated here.

First, *on military security,* both sides in the Korean conflict must:

- initiate a series of confidence-building measures to reduce the level of tension between the two sides, such as prior notification of the scheduled military exercises, sending an observer team to watch the exercises in progress, etc.;
- stop or reduce the frequent military exercises, such as the annual joint U.S.-ROK military exercises known as "Team Spirit," as a symbolic gesture of tension reduction;[46]
- agree to demilitarize the Joint Security Area in Panmunjom and to an eventual re-demilitarization of the DMZ.

Second, *on inter-Korean dialogue,* the two regimes must:

- sustain the current momentum of the North-South Korea dialogue (interrupted momentarily) at three levels: Red Cross talks, economic talks, and parliamentarian talks, so as to reach an agreement on substantive issues pertaining to tension reduction;
- take bold initiatives to resolve the current deadlock on sports talks to enable the holding of the 1988 Seoul Olympics as scheduled;*
- hold a summit meeting between Chun Doo Hwan and Kim Il Sung aimed at a breakthrough in overcoming the current deadlock in relations.†

Third, *on the diplomatic level,* both states must engage the major powers surrounding the Korean peninsula to assist in the resolu-

---

*Fortunately, there are indications that Pyongyang has agreed, in principle, to accept the International Olympic Committee recommended formula of allotting numbers of games to North Korea (out of a total of 23 games), although the details remain to be worked out.

†The possibility of Chun-Kim summitry cannot be ruled out completely, although the time is running out before President Chun's announced step-down from office in February 1988.

tion of the Korean conflict by taking any or all of the following measures:

- hold an international conference on the future of Korea;
- encourage diplomatic cross-contacts and recognition by the major powers;
- assist both Koreas in agreeing on measures of arms control and disarmament.

Fourth, and finally, *on the strategic level*, the major powers surrounding the Korean peninsula must agree among themselves and move toward adopting any of the following measures for peace on Korea:

- stop Korea's proxy or client role abroad;
- reduce Korea's arms sales abroad;
- denuclearize the Korean peninsula and proclaim Korea a nuclear-free zone.

V. Conclusion

Inter-Korean relations, especially as seen in the Korean arms race, have gone through the familiar pattern of action and reaction. Inter-Korean competition has intensified in the political, economic, and military arenas, with different sequences of development. In the 1970s, with the initiation of the North-South Korean dialogue on unification, the two states became aware of each other's strengths and weaknesses through an initial direct encounter in 1972–73, and subsequently through the resumed negotiation in 1984–85.

Inter-Korean competition has thus acquired a new intensity and seriousness of purpose in the 1980s. New elements of sophistication and complexity have been added, as both states are adopting new pragmatic postures and flexible styles of negotiation to influence the public opinion within and outside Korea. In spite of these encouraging recent developments, the fact remains that the Korean peninsula continues to be an explosive powder keg, where a slight provocation might engulf both states in flames. The outside powers which maintain alliance relations with each Korea, and an active interest in the peninsula, will surely be drawn into the fire in the process.

The Korean peninsula technically still remains a war zone because no peace treaty was signed to terminate the Korean War. Only an Armistice Agreement was signed on July 28, 1953, placing both sides under an armistice regime, an arrangement which is precarious and fragile at best. The Korean peninsula in the late 1980s is thus in a military stalemate, although technology drives the lethality of each side's forces ever upward. The area has turned into an armed camp, with the deployment of more than 1.5 million troops confronting each other across the narrow strip of the DMZ, 2.484 miles (4 kilometers) wide and 155 miles long, that separates the two sides militarily. This abnormality has been allowed to last for too long, for more than thirty-seven years since the Korean War broke out in 1950. It must be put to an end.

Since a divided Korea continues to remain a potential powder keg, or volatile flashpoint, positive measures for resolving the conflict need to be explored, lest an armed conflagration in Korea involve all the concerned parties, including the major powers with interests in the region of Northeast Asia. Thus, the future stability and deescalation of conflict on the Korean peninsula will be in the best mutual interest of all the parties concerned. The resolution of the conflict will depend on the outcome of the interplay of three dynamic factors: (1) inter-Korean relations, especially the political, economic, and military developments between the two states; (2) the major power policies, and especially the alliance relationship between the two regimes and their respective allies; and (3) the domestic political situation of the respective states, which may spill over into the external environment in the region.

As a divided Korea was the microcosm reflecting a world divided in the past, so the search for peaceful resolution of the Korean conflict may provide an opportunity, or an inspiration in the future, for a lasting world peace "without war and conflict."

This dream may sound idealistic. Yet, without it, no creative action is possible in the search for such a noble cause as institutionalizing the peace process on the Korean peninsula.

Notes

1. Young Whan Kihl, *Politics and Policies in Divided Korea: Regimes in Contest* (Boulder, Colo.: Westview Press, 1984).

2. Frederick Nelson, *Korea and the Old Orders in Eastern Asia* (Baton Rouge, La.: Louisiana State University Press, 1946).
3. Zbigniew Brezezinski, *Game Plan: A Geostrategic Framework for the Conduct of the U.S.-Soviet Contest* (New York: Atlantic Monthly Press, 1986).
4. See, for instance, President Ronald Reagan's press conference of 13 February 1986 in reference to the Philippine situation, regarding some of these chokepoints to contain the Soviet expansionism.
5. Joseph A. Yaeger, "The Security Environment of the Korean Peninsula in the 1980s," *Asian Perspective*, Vol. 8, No. 1 (Spring–Summer 1984), p. 86.
6. Richard L. Sneider, "Prospects for Korean Security," in Richard H. Solomon, ed., *Asian Security in the 1980s: Problems and Policies for a Time of Transition* (Santa Monica, Ca.: Rand, 1979), p. 112.
7. T. B. Millar, "Introduction: Asia in the Global Balance," in Donald Hugh McMillen, ed., *Asian Perspectives on International Security* (London: Macmillan, 1984), pp.5–6. Italics added.
8. Norman D. Levin, The Strategic *Environment in East Asia and U.S.-Korean Security Relations in the 1980s* (Santa Monica, Ca.: Rand, March 1983). See also Young-Koo Cha, "Strategic Environment of Northeast Asia: A Korean Perspective," *Korea & World Affairs*, Vol. 10, No. 2 (Summer 1986), pp. 278–301.
9. Levin, *The Strategic Environment in East Asia*, p. v. On a recent study of the Soviet nuclear buildup in Asia, see Richard H. Solomon and Masataka Kosaka, eds., *The Soviet Far East Military Buildup: Nuclear Dilemmas and Asian Security* (Dover, Mass.: Auburn House Publishing Co., 1986).
10. Levin, *The Strategic Environment in East Asia*, p. v.
11. A dependency that is permitting the Soviet Union considerably greater access to North Korean naval ports and air space. See *Christian Science Monitor*, 10 November 1986, p. 18.
12. Young Whan Kihl, "The 5th Column: Korea's North-South Dialogue Rests on a Powder Keg," *Far Eastern Economic Review*, 17 October 1985, pp. 44–45.
13. See, for instance, Young Whan Kihl, "North Korea in 1984: The Hermit Kingdom Turns Outward!", *Asian Survey*, Vol. 25, No. 1 (January 1985), pp. 65–79; Young Whan Kihl, "North Korea's New Pragmatism," *Current History*, Vol. 85, No. 510 (April 1986), pp. 164–167, 198.
14. *Yomiuri Shimbun* (Tokyo), 17 September 1985; *The Asian Wall Street Journal*, 21 October 1984; *The Washington Times*, 26 November 1985.
15. On the official denial by Seoul and Pyongyang, respectively, see *The Korea Times*, Chicago edition, 2 January 1986.

16. Young Whan Kihl, "Strategies for Improving Inter-Korean Political Relations," *Korea Observer*, Vol. 17, No. 4 (Winter 1986), pp. 382–400.
17. *Christian Science Monitor*, 11 February 1986, p. 11.
18. *The Military Balance, 1986–87* (London: IIS, 1986), pp. 159–161. See also Larry A. Niksch, "The Military Balance on the Korean Peninsula," *Korea & World Affairs*, Vol. 10, No. 2 (Summer 1986), pp. 253–277.
19. *The Military Balance, op. cit.* See also U.S. Arms Control and Disarmament Agency, *World Military Expenditures and Arms Transfers: 1971–1980* (Washington, D.C., 1983).
20. Kihl, *Politics and Policies in Divided Korea*, pp. 146–147.
21. *Ibid.*, p. 147.
22. This is a statement made by Admiral Ronald Hays, Commander-in-Chief of the U.S. Pacific Command, at a news conference in Bangkok, Thailand, in February 1986. See *The Korea Herald*, 7 March 1986.
23. On the relative evenness of the North-South Korean forces, see *The Military Balance, op. cit.* According to this source, "the opposing forces in the Korean Peninsula are roughly equivalent" (p. 118). See also Young Whan Kihl, "The Two Koreas: Security, Diplomacy and Peace," in Young Whan Kihl and Lawrence Grinter, eds., *Asian-Pacific Security: Emerging Challenges and Responses* (Boulder, Colo.: Lynne Rienner Publishers, 1986).
24. *A Comparative Study of the South and North Korean Economies* (Seoul: National Unification Board, 1985).
25. An estimate of the GNP figures for both Koreas was derived from the various issues of the *World Bank Atlas* and *Asia Yearbook*. Also consulted were: Kihwan Kim, *The Korean Economy: Past Performance, Current Reforms, and Future Prospects* (Seoul: Korea Development Institute, 1985), and *A Comparative Study of the South and North Korean Economies, op. cit.*
26. This is the conclusion arrived at in a U.S. CIA study, *Korea: The Economic Race Between the North and the South: A Research Paper* (Washington, D.C.: National Foreign Assessment Center and the Library of Congress, 1978), p. 2.
27. *Ibid.* p. 6
28. *Ibid.*
29. *Ibid.*
30. *Ibid.*
31. U.S. Congress, Senate Committee on Foreign Relations, *U.S. Troop Withdrawal from the Republic of Korea*, A Report to the Committee by Senators Hubert H. Humphrey and John Glenn, 9 January 1978, 95th Congress, 2d Session (cited hereafter as Humphrey-Glenn Report;

Washington, D.C.: GPO, 1978); Ralph N. Clough, *Deterrence and Defense in Korea: The Role of U.S. Forces* (Washington, D.C.: Brookings Institution, 1976).
32. Young-Ho Lee, "Military Balance and Peace in the Korean Peninsula," *Asian Survey*, Vol. 21, No. 8 (August 1981), pp. 852–864. See also Niksch, "The Military Balance," *op. cit.*
33. U.S. Congress, Humphrey-Glenn Report, pp. 35, 39.
34. *Ibid.*, p. 40.
35. For instance, see U.S. CIA, *Korea: The Economic Race.*
36. Young Choi, "The North Korean Military Buildup and Its Impacts on North Korean Military Strategy in the 1980s," *Asian Survey*, Vol. 25, No. 3 (March 1985), pp. 343–348.
37. *Ibid.*, p. 355.
38. *Ibid.*
39. *Japan Military Review* (February 1986), as reported in *The Korea Herald*, 18 February 1986.
40. Solomon, *Asian Security*, p. 31.
41. Richard L. Sneider, *The Political and Social Capabilities of North and South Korea for the Long-Term Military Competition* (Santa Monica, Ca.: Rand, January 1985), pp. 35–36. The author of this report was former U.S. Ambassador to ROK during the Ford administration.
42. *Ibid.*
43. *Ibid.*, p. vi.
44. *Ibid.*, p. 44.
45. *Ibid.*, pp. vi–vii.
46. A Soviet-North Korean naval exercise, the first of its kind, was reportedly held in October 1986 in the Sea of Japan.

6. The South China Sea: From Zone of Conflict to Zone of Peace?

Donald E. Weatherbee

THE geographic core of Southeast Asia is the maritime zone of the South China Sea. Its littoral states comprise the Philippines, Malaysia, Brunei, Indonesia, Singapore, and Thailand—thus, the six states of the Association of Southeast Asian Nations (ASEAN); the Indochinese states of Kampuchea (Cambodia) and Vietnam; and the two Chinas—the People's Republic of China (PRC) and Taiwan. Major maritime access to or egress from the South China Sea basin is at the north through the relatively narrow passages of the Bashi Channel or Formosa Straits, and in the south through the even narrower and territorialized Malacca and Sunda straits.[1] To this area we reckon the Gulf of Tonkin, the Gulf of Thailand, and, as an arm, the Sulu Sea, entered through the Balabac Strait.

Calling attention to the South China Sea's almost land-locked nature (90 percent of its circumference rimmed by land), one regional political geographer has conceptualized it as a "geopolitical lake," over which competitive claims to territory, maritime and seabed jurisdictions, and fisheries bring the littoral states into a complex web of conflict and rivalries.[2] The most important resource at stake is energy in the form of proven offshore oil and natural gas reserves.[3] For the time being, at least, the resource stimulus for much of the competition has been relegated to the background as oil prices plunged and consumption patterns changed. Because of the South China Sea's status as a "semi-enclosed sea" under the new Law of the Sea Treaty, the littoral states are encouraged to "cooperate with each other in the exercise of their rights and in the performance of their duties."[4] Even in the absence of bilateral local disputes, resting essentially on the interpretation of law and facts of history, the possibility for such "cooperation" tending toward making the South China Sea a zone of peace would be limited by the presence of other, even deeper political divisions, both in the region itself and linking the regional states to extra-regional actors.

Overarching the local and regional rivalries and competitions for territory, rights, and resources are the South China Sea's geostrategic functions.[5] It continues to play its historic role of a well-traveled thoroughfare for commercial traffic, linking East Asia to the Indian Ocean and beyond. In recent years this function has been overshadowed by its dimensions as a regional theater of great power confrontation. Both the United States and Soviet navies are present in force, with their basing respectively in the Philippines and Vietnam, but deploying from and through the South China Sea for extra-regional missions. Moreover, as China's naval modernization eventually gives it the ability to move from coastal defense to "blue water" deployment, its South Sea Fleet too could become a significant factor in the regional naval balance.[6]

The great power political/military presence in the region is functionally linked by both formal alliance and informal orientations to regional actors, and by extension regional disputes. The most significant of these regional conflicts is the Kampuchea issue. Therefore, in cases where competitive claims in the South China Sea are embedded in the context of deeper political antagonisms or concerns, as between Vietnam and China or Vietnam and the ASEAN states, a framework for peaceful resolution is difficult to negotiate. Even where there is an established structure for peaceful settlement, as in the ASEAN regime, some conflicts still prove intractable as, for example, the Philippine claim to Sabah.

Fortunately for regional stability, to date the local political competitions have not erupted into major armed conflict between contending claimants. Nevertheless, the possibility for serious confrontation is there and is being heightened by the developing regional mini-arms race leading to augmented and high-technology naval surface and air forces for all of the littoral states.[7] The potentially destabilizing impact of this arms race is dampened not only by the global change in petroleum supply and demand, but also by the superimposition of the great power balance on the regional balance, partially serving in a deterrent way to inhibit local adventurism. If events in the Philippines, however, should lead to a degradation of the U.S. capabilities in the region, the impact will be felt not just in the U.S.-Soviet power relationship but also in regional perceptions of disequilibrium, and a new, and to ASEAN, disadvantageous distribution of power in the South China Sea.

While the maritime boundaries in the South China Sea have historically been subject to the ebbs and flows of littoral states' political and commercial interests and power, it has only been in recent years that the existence of competitive jurisdictional claims has received priority attention in the economic, political, and security considerations of the littoral states and the extra-regional users. The geopolitical foundation of the competition is to be found in disputed territorial claims to islands and sea space, as well as in the mosaic of overlapping jurisdictions in the new maritime Exclusive Economic Zones (EEZs) and Continental Shelf zones, in addition to some special purpose—for example, security—sea and air space zones.[8]

I. Conflicting Claims

China claims all of the more than 200 islands, reefs, and shoals in the South China Sea. These are divided into four groups: in the northeast, the Dongsha Islands (Pratas); in the west, the Xisha Islands (Paracel Islands); flanking the Paracels in the east and southeast, Zhongsha (the Macclesfield Bank) and Huangyen (Scarborough Shoal); and the Nansha (Spratly) Islands in the south. Zhou Enlai's commentary on the retrocession terms of the 1951 Japanese Peace Treaty echoed the established position of the preceding Kuomingtang government with respect to the fact that these islands "have always been Chinese territory."[9] China's claims to sovereignty over the islands "since very ancient times" are based on assertions about discovery, development, and continuous administration and jurisdiction that prove that "the South Sea islands have always been part of China's territory" and that "the Chinese people" have "indisputable sovereignty." As part of

China's "sacred territory," it is China's duty to defend them and where alienated to liberate them.[10] From these territorial points, the total expanse of China's maritime territorial claims in the South China Sea lies like a giant overlay over the jigsaw puzzle of the claims of the other contending littoral states, extending in the south to 4 degrees North latitude at Malaysia's Tseng-mu Reef, only 20 miles off the Sarawak coast. In its full extent, the Chinese claim would cut off the other littoral states from the energy resources of the South China Sea. The seriousness of China's South China Sea claims are recognized by the regional states. As one leading Indonesian analyst put it: "It is expected that China will maintain and defend the [South China Sea] claim, perhaps by force of arms if necessary."[11]

Dongsha (or Pratas Reef), where Chinese sovereignty is not contested, is currently occupied by Taiwan. The shoals and reefs of Zhongsha (or the Macclesfield Bank) do not emerge even at low water. It does, however, serve as a mid-sea anchorage area. The Chinese (and Taiwanese) assumption of territoriality—even to the extent of naming the submerged formations—provides a reference point for contiguous water and seabed claims. It is really in the Paracels and the Spratlys where China confronts Vietnam, and supposedly its Soviet ally, that potential flashpoints for war are clearest.[12] The mix of motives driving national policy includes political, strategic, and economic considerations. As Selig Harrison has argued for China, for example: "To put the oil factor into a meaningful perspective, it should be viewed as one element in a more comprehensive Chinese effort to consolidate a position of regional primacy."[13]

In the "100 minute war" of January 16, 1974, during the waning of the Saigon regime, the Chinese drove the South Vietnamese garrison out of the Paracels (Vietnamese: Hoang Sa) annexed by France in 1932. The Paracels are a group of fifteen islets together with a number of reefs, and shoals scattered over a rough oval 125 miles long in the middle of the Gulf of Tonkin, with its center about 220 miles south of Hainan and 250 miles east of Danang. The Socialist Republic of Vietnam's successor government has argued its claims to sovereignty in vigorous verbal statements and "White Books."[14] It has not, however, tried to physically challenge what seems to be a rather light Chinese presence there. Neither has China sought to take strategic advantage from its position. For example, during the 1979 Chinese "punishment" of Vietnam's invasion of Kampuchea, there was no activity in the Paracels. Intervening between China and the Vietnam, of course, is the Soviet Union and its naval capabilities. This must act as a deterrent to any maritime-based Chinese military actions against Vietnam. On the other hand, while Moscow decries Chinese "aggression" against Vietnam, it has not lent Hanoi naval assistance to recover its position in the Paracels. Obviously, for Moscow unrestrained Vietnamese nationalism will not be allowed to be a *casus belli* in escalating the Sino-Soviet conflict.

The situation in the Spratlys is different, being more complicated and militarized. The Spratly archipelago is the largest of the South China Sea island groups, consisting of thirty-three islands, shoals, reefs and banks above the low water mark. Lying roughly in the middle of the 594-mile-wide South China Sea, the Spratlys stretch out over 70,312 square miles. With the Paracels, the Spratlys straddle the strategic sealanes of Southeast Asia. To claim, however, that control over these islands would give the controlling nation a stranglehold on Japan or a veto on the strategic movements of the great powers seems to exaggerate the strategic importance of what in fact are specks of land. Of the five outcroppings that can be dignified by the name island, the largest, Thithu (or Pagasa), occupied by the Philippines, is less than 1 mile long by 625 yards wide. Itu Aba, where there was once a Japanese seaplane base and now the site of a Taiwanese "naval base," is 960 yards long by 400 yards wide. The Vietnamese "base" is situated on Spratly Island (Truong Sa), which is barely over 700 yards long.

China lays claim to the Spratlys. It has not, however, made any attempt to seize the archipelago, all of which is under the control of other governments. France had occupied nine of the Spratlys and in 1956 the South Vietnamese government began to garrison the islands, annexing them to Phuoc Tuy Province in 1973. The new Vietnamese regime hastened to make its presence felt by dispatching forces which now occupy six islands and atolls in the west and central part of the archipelago. Manila proclaimed its sovereign rights in what it terms the Kalayaan (Freedom) Islands in 1955, but did not begin to occupy the northern Spratlys until 1968. It garrisoned three islands: Pagasa, Nanshan, and Flat. Manila's territorial claim was closely associated with the prospective offshore oil field at the Reed Bank. Although the Philippine territorial claims in the Spratlys are legally tenuous,[15] it does not appear that the Aquino government intends to relinquish them, continuing to maintain that the Kalayaan Islands are separate from the Spratly Islands.

In the early 1980s, both Vietnam and Manila bolstered their military forces in the archipelago. Vietnam has forces on four islands with hardened defensive positions and a surfaced runway on Spratly Island (Truong Sa). According to Hanoi, since the Spratly "liberation" on April 16, 1975, the Vietnamese forces in the islands have continuously enhanced their position, "in order to stand constantly ready to firmly defend the fatherland's territorial waters."[16] In 1976, Manila created a Western Command headquartered at Puerto Princesa on Palawan, along with a $150 million buildup for the defense of the Spratlys.[17] In 1978, Philippine marines landed on the remote sandbar of Panata, the last unoccupied usable piece of land. In 1982, Prime Minister Cesar Virata toured the Kalayaan Island's fortifications and declared: "We will defend the Kalayaan because it is ours. Any offensive action against Kalayaan will be considered as an assault on the sovereignty of the Republic of the Philippines."[18]

The Philippines has tried to take the problem of jurisdiction in the Spratlys out of the local, bilateral context and raise it as a wider regional strategic problem. For example, in a March 1983 Defense Ministry report, the strategic threat to ASEAN from an unfriendly power based in the Spratlys was mooted.[19] Furthermore, the report went on to warn that after proper mapping, the Spratlys could be a launching area for ballistic missile–capable submarines. The possible linkage between the local contending powers in the Spratlys and their great power allies is ambiguous. If in the event of an attack on the Philippines armed forces in the islands, for instance, Manila were to invoke the U.S.–Philippines Mutual Defense Treaty, it does not technically apply since the Spratly Islands, occupied only in the past two decades, lie outside the Spanish-American treaty limits defining Philippines' sovereign territory. On the other hand, U.S. interests would not be unaffected if hegemony by a hostile power were to be exercised over the archipelago. Similarly, while the USSR has not directly assisted Vietnam to assert its territorial claims in the South China Sea, its deterrent presence to those who might challenge the status quo established by Vietnam in the Spratlys cannot be ignored. This applies in particular to China. A brief alarm sounded in May 1983, when China conducted a naval exercise in the vicinity of the Spratlys that included deployment of amphibious landing vehicles with troops, but no landings were made.[20] The Chinese show of force might be interpreted as a response to the April 1983 Soviet amphibious exercise, which included the landing of naval infantrymen from the *Ivan Rogov* on the Vietnamese coast.

Malaysia raised the stakes in the South China Sea territorial game, as well as complicating geostrategically its own defense planning, when it decided to enter the competition in the Spratlys in earnest. In June 1983, in the operational framework of its first full-scale combined military exercise in its claimed maritime zone, Malaysian amphibious forces occupied Terumbu Layang-Layang in the southeastern corner of the island group. This action was immediately protested by China, which claimed "indisputable sovereignty" over the reef it calls Danwan,[21] as well as by Vietnam.[22] The military exercise that provided the cover for the occupation of Terumbu Layang-Layang was intended to demonstrate that Kuala Lumpur was serious about defending its territorial claims, which extend in the Spratlys to include Amboyna Key (Malaysia: Pulau Kecil Amboyna), some 64 kilometers northwest of Terumbu Layang-Layang, which has been occupied by Vietnam since 1978. Malaysia then rejected the Vietnamese position, stressing that it took "a serious view of the stationing of troops and strongly protested against Vietnam's actions."[23] Malaysia claims both islands fall within its 200-nautical-mile Continental Shelf and EEZ boundary as published in its Continental Shelf Act of 1967 and shown on the map gazetted in December 1979 defining its maritime limits.[24] Its military action might be interpreted as preemptive—to keep the Vietnamese off Terumbu Layang-Layang.

Malaysia's decision to inject its power in the region and to use force if necessary to resist aggression (i.e., prevent expulsion) in its claimed jurisdictions gives new strategic importance to the island of Labuan, once the site of British naval presence in Borneo, and now the support base for Malaysian military activities in the middle reaches of the South China Sea. Priorities given to the upgrading of the naval and air facilities on Labuan in the defense budgets from 1979 to 1982 heralded its role as the eastern bastion for secure sealanes between peninsula Malaysia and Sarawak and Sabah, as well as a base for operations in the South China Sea.[25] Its importance was further underlined by the April 1984 cession of the 91-square-mile island by the Sabah state to the central government as a Federal Territory.

The enhanced Malaysian military presence on Labuan is not just relevant to its maritime zones. The island commands the entrance to Brunei Bay, and the new federal position may not be wholly unconnected to unresolved territorial and jurisdictional disputes with newly independent Brunei. More darkly, Sabah's chief minister at the time, the Berjaya Party's Datuk Harris Saleh, alluding to past plots to take Sabah out of the federation, stated: "It is for the purpose of deterring any similar attempts in the future that the presence of the federal government here is necessary."[26] What is described as an "immovable federal presence" is seen as a requirement of integration of the Borneo state into the Malaysian Federation. The terms of transfer, however, have been questioned by the successor Sabah government, the dominant party of which, Parti Bersatu Sabah, has called for a referendum on the return of Labuan to Sabah.[27]

The discussion of Sabah and its defense brings to the fore another unresolved territorial issue in the South China Sea region: the Philippine claim to Sabah. The Philippine claim to sovereignty over Malaysia's North Borneo, Sabah state was initially advanced in 1962, and while not actively pursued in recent years, still after a quarter of a century irritates normal relationships between Malaysia and the Philippines.[28] The Philippines' national territory has been constitutionally defined as "territories belonging to the Philippines by historic right or legal title" and has been judicially understood to include Sabah. The Philippine Base Line Act of 1968 is much more specific. Section 2 states that: "The territorial sea of the Philippines as provided in this act is without prejudice to the delineation of the baseline around the territory of Sabah, situated in North Borneo, over which the Republic of the Philippines has acquired dominion and sovereignty." President Marcos's verbal renunciation of the Philippines' claim at the 1977 Kuala Lumpur ASEAN summit was never followed up by constitutional amendment or by statute in Manila. The international dispute over sovereignty has been complicated by the flow of Muslim Filipino refugees to Sabah from the Philippines' southern islands during the Moro insurgency and by Manila's suspicions that Sabah was being used as a sanctuary for the Moro National Liberation Front (MNLF).[29] To this mix must be added the "unofficial trade" carried out by well-armed smugglers and gun runners and the depredation of pirates operating in the poorly patrolled waters separating Malaysia and the Philippines' Sulu Archipelago.

The ASEAN framework has so far allowed Manila and Kuala Lumpur to localize flareups in the political/military tinderbox at their mutual peripheries. For example, in September 1985, Philippine-based pirates raided the Sabah east coast town of Lahad Datu. A week later local reports had it that the Malaysian navy had attacked in the Tawi-Tawi group in the Sulu Archipelago. Cooler heads in both Manila and Kuala Lumpur quickly moved to limit the political damage done by what Malaysia called a "fabrication," and acknowledged by senior Filipino officials as a "provocation" by forces seeking to disrupt relations between the two countries.

The position of the Aquino government has been ambiguous. While its spokespersons recognize that settlement of the issue is necessary if relations with Malaysia are to be normalized, what is seen as the domestic exigencies of Muslim politics have been given great weight. The October 1986 draft constitution did not terminate the claim. Mrs. Aquino herself has said that she wants to settle the claim once and for all on the basis of self-determination and justice. This will not be satisfactory to Kuala Lumpur, which feels that self-determination and justice are already served in the federation. It appears that an eventual resolution of the issue will be consequent to a settlement of the MNLF's war in the Philippines and the Filipino refugee problem in Sabah. Until then the issue will continue to bedevil Philippines-Malaysian bilateral relations and impede progress in ASEAN's patterns of multilateral cooperation. The long-awaited third ASEAN summit that was scheduled for mid-1987 in Manila was within three months of announcement pushed back to December 1987, in part because of Malaysia's long-standing reluctance to grace Manila with high-level visits.

II. Economic Aspects

The economic dimensions of the competitive claims in the South China Sea are as important as the political and strategic implications. This is particularly true as the new Law of the Sea Convention prescribes new legal rights as well as duties on the littoral states as well as other users of the maritime space. The convention accepted the de facto state practice of a 12-nautical-mile territorial sea breadth. Beyond that, the new Law of the Sea accepts the 200-nautical-mile Exclusive Economic Zone within which the state has the right to exploit and manage the living and nonliving resources (Articles 2 and 3). The question of jurisdiction over nonliving resources raises the issue of overlapping Continental Shelf zones which extend beyond the territorial subsea to the outer edge of the continental margin, or to a distance of 200 nautical miles, whichever is further (Article 76). For Southeast Asia, this means the effective Continental Shelf limit for any littoral state is 200 nautical miles from its territorial base line, coextensive, therefore, with the EEZs and having the same kinds of overlaps.[30] In this zone the coastal state has the exclusive rights of jurisdiction and resource exploitation. The multiplicity of self-proclaimed EEZs and Continental Shelf limits in the South China Sea has created new kinds of territorial issues that are now part of the politics of changing power relationships in the region.[31] The first question is from where and how the territorial baselines are drawn, outwards from which the other zones are measured. The second question is where the zones overlap, how the boundaries are to be delimited.

It is obvious that sovereignty over the disputed islands in the South China Sea would carry with it substantial extensions of a nation's EEZ and Continental Shelf, and thus control over fisheries and under seabed resources. Another type of territorial claim, however, is that encompassed in the "archipelagic principle" now accepted in the Law of the Sea Convention (Articles 46–53). For Indonesia and the Philippines, the promoters of this principle, the inherent and intrinsic unity of the archipelagic state is given legal effect by drawing its territorial baselines in a way to connect the outermost points of the outermost islands. It is from these baselines that the territorial sea and other contiguous zones are measured.[32]

In the South China Sea, the Indonesian territorial baseline runs from Tanjung Datu in West Kalimantan, northwest to Natuna Utara (approximately 5 degrees North latitude and 108 degrees East longitude), southwest to Anambas, and then to the northern tip of Bintan on the Singapore Straits. This closes the southern end of the South China Sea and cuts the direct route between Peninsula Malaysia and East Malaysia. The major impact of the Philippines' baselines is to give legal weight to Manila's assertion of sovereignty over the Sulu Sea, with the territorial line being drawn southeastward from Cape Melville on Balabac to the midpoint of the Sibulu Passage between Sabah and the Tawitawi group.

The jurisdictional enclosure of the archipelagic waters, done in Indonesia by unilateral declaration and statutory act and in the Philippines by constitutional provision, now is enshrined in the new Law of the Sea Convention. The waters inside of the archipelagic baselines become internal seas with appertaining rights. Although freedom of transit is guaranteed, these are no longer high seas, and the controlling state may designate navigational routes. Furthermore, passage must not be prejudicial to peace, good order, or the security of the archipelagic state. Moreover, regulations of navigation can be made for purposes of safety, conservation, and enviromental protection. The conditional nature of navigation, therefore, means the archipelagic state can broadly interpret its obligations to maintain freedom of transit.

While both Indonesia and the Philippines can make compelling arguments that the archipelagic baselines are necessary for the security and territorial integrity of the state, creating thereby an indissoluble unity of land, water, and population,[33] the economic impact in terms of jurisdiction over resources is also important. This is very clear when we note Indonesia's 1980 proclamation of its EEZ that carries its resource boundaries deep into the South China Sea from its archipelagic baselines.[34] As far as Indonesia is concerned, its EEZ is part of the unity of land and water over which it has sovereignty. The proclamation added 976,600 square miles to its jurisdiction which, according to the Indonesian foreign minister, existed with or without a Law of the Sea Convention.[35]

The country most deeply affected by Indonesia's maritime zones is Malaysia, which proclaimed its own EEZ in May 1980, overlapping all of its ASEAN neighbors and Vietnam. Malaysia acknowledges Indonesia's archipelago principle, in return for which a 1982 treaty between the two countries defines Malaysian rights of communication access between Peninsula Malaya and East Malaysia. Indonesia's consistent approach to all of its overlapping areas has been one of seeking friendly negotiations and delimitation on the basis of an agreed midline. With respect to other overlapping jurisdictions between ASEAN countries, Indonesia and Malaysia (1969) and Indonesia and Thailand (1971) have formally agreed to delimitations of their Continental Shelf jurisdictions.[36] A tripartite agreement in 1971 between Malaysia, Indonesia, and Thailand delimited the Continental Shelf boundaries in the northern end of the Straits of Malacca. Malaysia and Thailand have agreed in principle to joint exploitation of oil and gas in their overlapping zones in the Gulf of Thailand. A February 1979 understanding provides for a joint administration of the overlap.[37] The joint authority's constitution was to be given the "force of law" once the technical details were worked out. Implementation has been delayed because of differences in the legal systems of the two countries. However, political issues may also be intervening.

The Philippines has not yet delimited its zones with Malaysia or Indonesia because of the political obstacles of other jurisdictional disputes. As already noted, the Sabah question controls Malaysia-Philippines relations. Until this is satisfactorily resolved, the EEZ overlaps which have caused friction in the fisheries contiguous to Commodore Reef will persist. For Indonesia, it is the continued Philippines' claim on its "historic territorial waters" around the Indonesian island of Miangas (Palmas) that is the obstacle to boundary delimitation.[38] Until the territorial question is settled, the other EEZ and Continental Shelf jurisdictional overlaps between Mindanao and the Indonesian Sangihe and Talaud island groups will remain undemarcated.[39]

A Continental Shelf issue exists between Malaysia and Brunei. By a proclamation in 1954, Brunei annexed its Continental Shelf to a depth of 595 feet, adding thereby 5,700 square miles to the state's resource base. This was ignored by Malaysia when it gazetted its East Malaysian Continental Shelf limits in December 1979. The shelf was shown as uninterrupted from the Indonesian border to the Philippines, with no shelf being shown for Brunei. Malaysia refused to negotiate with London on the matter as part of its tactic of pressing for rapid decolonization of Brunei. The Malaysian negotiating position now that Brunei is independent may not be unconnected to the continued assertion by Brunei of sovereignty in the so-called Limbang Salient, part of the East Malaysian Sarawak state, the last bite taken from Brunei by the Brooke's Sarawak in 1890. Although the British protecting power hesitated to impinge on the welter of competitive claims and disputed jurisdictions in the South China Sea, an independent Brunei followed the lead of the other littoral states and in late 1983 proclaimed its 200-nautical-mile EEZ, quadrupling thereby its maritime jurisdictions, but creating a whole new set of multiple overlaps.

The ability of the ASEAN states to settle their boundary issues cooperatively by negotiation and ASEAN's common interest in a peaceful maritime regime has led Dr. Phipat Tangsubkul, Thailand's leading academic expert on Law of the Sea questions, to propose that ASEAN as a grouping declare an EEZ for the whole of ASEAN as a way to avoid intra-ASEAN disputes.[40] Phipat's suggestion reflects Thailand's relatively disadvantaged EEZ position (declared in 1981) in the Gulf of Thailand and the Andaman Sea. Malaysia's vigorous implementation of its fisheries jurisdiction in its EEZ, once traditional Thai fishing areas, has created political headaches for Bangkok.

Vietnam's territorial sea claims were promulgated by its National Assembly on November 12, 1982, drawing the baselines from which other maritime zones are measured from the Vietnamese mainland to its claimed island sovereignties in the Gulf of Tonkin and the South China Sea.[41] The "statement," which codifies earlier National Assembly declarations, is premised on the continued legal validity of the 1887 French-Chinese Gulf of Tonkin territorial demarcation, which for China today represents part of the web of inequalities imposed by imperialism. Not surprisingly, the Chinese reaction was to declare the claimed boundaries "null and void" as far as China was concerned, warning that with respect to its "expansionist designs," "the Vietnamese authorities must bear full responsibility for all the serious consequences that might arise therefrom."[42] In the Tonkin Gulf (Vietnam: Bac Bo; China: Beibu), Vietnam's claim of a 108th meridian boundary would put nearly two thirds of the Gulf (and resources) under Hanoi's control.[43]

Territorial and maritime zone disputes which had troubled relations between South Vietnam and Cambodia and then the Socialist Republic of Vietnam and Democratic Kampuchea have been resolved between the SRV and the People's Republic of Kampuchea (PRK). In May 1975, Vietnam (through the Provisional Government of the Republic of South Vietnam) unilaterally altered its maritime border with Cambodia—the Brevie Line—by forcefully seizing the disputed islands of Phu Quoc and Puolo Wai, nearly doubling its Continental Shelf claim in the Gulf of Thailand and its overlap with Thailand while reducing Cambodia's by two thirds. Although Puolo Wai was later turned back, Vietnam still refused to accept the Brevie Line as the baseline for measuring its maritime zones.[44] The respective declarations of 200-nautical-mile EEZs by Vietnam (1977) and Democratic Kampuchea (1978) accentuated their differences. However, in July 1982, a pliant PRK reached an agreement with Vietnam on delimiting their "historical waters."[45]

Thailand, whose Continental Shelf and EEZ overlaps with both Kampuchea and Vietnam, refuses to recognize the expansive jurisdictional claims of Vietnam in the Gulf of Thailand. In a statement issued on November 22, 1985, the Thai Foreign Ministry categorically rejected Vietnam's November 1982 proclamation of its baselines as contradictory to the Law of the Sea and based on islands too far from land to serve as reference points. As for the PRK-SRV agreement, the Thai position is that it is "devoid of any legal effect" since the PRK government is not the legitimate government of Cambodia.[46] Thai insistence on reserving all of its rights is based not only on potential oil deposits in the overlap areas but on important Thai fisheries as well. Thai fishing boats are often seized in the disputed waters by Vietnamese patrol boats.

There remains a potentially explosive 7,800-square-mile overlap between Indonesia and Vietnam north of Indonesia's Natuna Islands, where Indonesian oil concessions have been granted on what the SRV claims is its Continental Shelf. A continuing series of bilateral negotiations on the demarcation question has gone on since 1978, but with no success. Ostensibly, the dispute is over the technical question of what method of demarcation should be used, with Indonesia insisting on the commonly accepted median line rule while Vietnam unorthodoxly presses for the application of the *thalweg* principle. This is normally used for delimiting international boundaries in rivers along the deepest parts of the river bed.[47] Hanoi claims that a trench running from the northern end of the Anambas to a region just north of the Natuna Islands marks the end of its Continental Shelf, and that the boundary should be along the deepest points of the claimed trench. Although Vietnam has declared that the issue of contested jurisdiction will not lead to armed conflict, it has tried to warn off foreign oil contractors from Indonesian-granted concessions in the disputed zone. Indonesia, for its part, has declared that it will protect the oil contractors—by force if necessary.

Indonesia's defense buildup in the late 1970s and early 1980s placed great strategic emphasis on developing capabilities to project power into its EEZ and to provide for a first line of defense in the air and sea space north of the Natuna Islands.[48] At the core of what ASEAN calls its "regional resilience" is an implicit strategic alliance between Indonesia and Malaysia.[49] The two countries have defined a common security interest in the defense of their South China Sea jurisdictions, with the Indonesian commitment to common defense extending to Terumbu Layang-Layang.[50] Bilateral exercises between the military forces of Indonesia and Malaysia with other ASEAN countries are becoming a regular feature in the South China Sea air and sea space. Furthermore, ASEAN exercises with friendly extra-regional military forces are not uncommon. In August 1986, the largest gathering of naval vessels for an exercise under the Five Power Defense Arrangement assembled in the South China Sea. A five-nation, twenty-six-ship task force maneuvered in an operation codenamed "Starfish." The United States deploys the Seventh Fleet regularly in the region. The "Cobra Gold" joint exercises with Thailand are an annual demonstration of the U.S. commitment to the security of the region.

Despite the shows of force, it does not appear likely that the local protagonists of the South China Sea imbroglios will readily turn to war to settle their disputes. As we have noted, Indonesia and Vietnam continue their dialogue even as Vietnam reserves its rights to the disputed overlap north of the Natunas Islands. Even though the Vietnam and the Philippines armed forces face one another in the Spratlys, both seem satisfied with merely holding on to the status quo rather than trying to dislodge each other. President Marcos claimed in 1978 that he had agreements with both China and Vietnam to settle any disputes in the South China Sea through negotiations.[51] In face-to-face discussions at the foreign ministerial level, both Vietnam and Malaysia have stressed the importance of settling conflicting claims in the South China Sea in a "friendly way."[52] Senior Malaysian officials have pointed out that neither side had made attempts to dislodge the other, suggesting that the status quo at Amboyna Cay and Terumbu Layang-Layang is acceptable.[53] As long as a rough great power balance operates in the region, it does not seem probable that either the USSR or China would attempt unilaterally to upset the status quo, particularly in the environment of the ongoing Sino-Soviet "normalization talks." Although there might be some uncertainties for the various oil concessionaires with respect to their position in disputed zones, the current economic climate for exploration and increased production does not suggest such urgency to resolve questions that the use of force is tempting.

III. The ASEAN Position

The inability of the ASEAN states and Vietnam to adjust their claims and delimit their jurisdictional boundaries seems less a matter of bilateral differences in the application of legal principles than the fact that the South China Sea disputes are caught up in the broader political struggle between ASEAN and Indochina over the future of Kampuchea. It has generally escaped notice

THE SOUTH CHINA SEA 139

that the maritime questions were included as point four of the 1981 Indochinese statement on "Principles on Relations Between Indochina and ASEAN," presented to the U.N. General Assembly.[54] Assuming an eventual post-Kampuchea crisis framework of peaceful relations between Indochina and ASEAN, a closer look at the text of this point is warranted:

> To respect the sovereignty of the coastal countries of the South China Sea over their territorial waters as well as their sovereign rights over their exclusive economic zones and continental shelves.
> To ensure favorable conditions for the land-locked countries in the region regarding the transit to and from the sea, jointly guarantee maritime rights and advantages to the same countries in accordance with international law and practice.
> To solve disputes among the coastal countries of the South China Sea over maritime zones and islands through negotiation. Pending a resolution, the parties concerned undertake to refrain from any actions that might aggravate the existing disputes. The various countries in the region will act jointly to seek modalities of cooperation among themselves and with other countries inside or outside the region in the exploitation of the sea and seabed resources on the basis of mutual respect, equality and mutual benefit, preservation of the environment against pollution, guarantee international communications and the freedom of the sea and air navigation in the region.

These principles could only become operative in the context of a modus vivendi for peaceful coexistence between Indochina and ASEAN. The international political basis for this symbolically exists in the notion of a Southeast Asian Zone of Peace, Freedom, and Neutrality (ZOPFAN). Although the tentative diplomatic gropings to make the Vietnamese and ASEAN concepts of a peace zone congruent were abruptly ended by the December 1978 Vietnamese invasion of Kampuchea, the ZOPFAN still has declaratory importance as the articulation of the ideal regional international order, which conceptually denies the permanent strategic division of Southeast Asia.

In 1984, the ZOPFAN reemerged as an important agenda item for ASEAN. The July 1984 Ministerial Meeting reiterated ASEAN's determination to work for a ZOPFAN and welcomed the revival of the ASEAN Working Group on ZOPFAN. This group, spurred by Malaysia, has agreed in principle that one of the first steps in a ZOPFAN would be to create a "nuclear weapons free zone" (NWFZ).[55] The quandary for planners of a Southeast Asian NWFZ is how to deal with U.S. basing in the Philippines and the USSR in Vietnam, let alone the question of air or sea transit of nuclear weapons-capable vessels. In other words, the critical question is great power strategic mobility. Indonesian Foreign Minister Mochtar elaborated on this at length in a Jakarta ZOPFAN seminar in early 1985.[56] He pointed out that it was no secret that nuclear-armed ships pass through Indonesian waters, thus affecting Indonesian security. A large part of the Indonesian discussion of operationalizing a NWFZ as the first step toward realizing a ZOPFAN is focused on the new international Law of the Sea. Jakarta places the political issue of a NWFZ in the context of its "archipelago principle" and the concomitant regulatory rights of the Indonesian state. In theory at least, with the declaration of a NWFZ, the transit of nuclear-armed ships could be declared prejudicial to public order and security and, hence, could be prohibited.[57]

Given the general global momentum toward nonnuclear regimes—in the Pacific, of course, the New Zealand case and the South Pacific Forum's Nuclear Free Zone Treaty—it is likely that ASEAN will move ahead with its effort to create a legal NWFZ. It will most probably be quite politically porous, grandfathering existing relationships and, like the South Pacific Forum, leaving the questions of transit and port calls to the individual regional states. Indonesia in particular, as long as it politically accepts the fact of the necessity of the strategic mobility of the United States as part of the maintenance of a regional balance of power, while reserving its legal rights, will not force the issue of U.S. navigation rights in the archipelago waters.

Although it is possible, as we have above, to place a discussion of Indonesian efforts to impose a restrictive transit regime on the naval super powers within an international law framework, for policy purposes such a discussion is fanciful, to say the least. In the last analysis, neither the United States or the USSR will be willing to accommodate what they perceive to be their own vital security interests to claimed territorial jurisdictions in maritime space. The real issues are not "rights" but how, if desired, such rights could be enforced, or what would be the international political conditions under which these rights would be respected. Neither a Southeast Asian ZOPFAN nor a NWFZ is a self-implementing regime. The great powers must accept it.

IV. Factoring in Great Power Rivalry

Great power strategic relationships in the South China Sea are not static. The so-called regional balance is dynamic, impacting on the way local powers perceive their own security requirements. Indonesia in particular is concerned about the strategic intentions of China and views with concern its increasing naval capabilities. Indonesian planners see the extension of China's naval reach to the South China Sea as a security threat to Indonesia and Southeast Asia, given both China's territorial claims and its historical claim to regional hegemony.[58] The fact that the U.S. Navy exercised with Chinese naval vessels in the South China Sea in late 1985 only underlines Jakarta's concerns about the regional implications of any U.S.-Chinese "strategic alliance."

The significance to be attached to the PRC as a great power in the South China Sea will be in part a function of what might be an even more significant alteration in the dynamics of the balance of power; that is, the relationship between the United States and the USSR. With the acquisition of basing facilities in Vietnam for its Pacific Fleet, the ability of the Soviet Union to project its power in and through Southeast Asia has been considerably enhanced. Of special concern is the steadily expanding threat to vital strategic sealanes and lines of communication running through Southeast Asia. A particular worry to China would be Moscow's willingness to support Vietnam's ambitions in the South China Sea. For the ASEAN states, any dramatic alteration of the status quo that would allow a potentially hostile power control in the region would be destabilizing, moving the perimeter of the strategic frontier between the ASEAN states and Communist states out to a sea zone in which Indonesia's archipelagic baselines become the front line, with the Natuna Islands as its "bulwark."[59]

In the intermediate-range future, the most important factor in determining the structure of the great power regional balance will be the fate of the American military installations in the Philippines. New basing negotiations are to commence in 1988, and the current agreement expires in 1991. There are deep and politically intense anti-base currents in the Philippines' domestic political scene. While the anti-base forces were not able to write their position into the constitution, the debate over the future of the bases will be polarizing and its outcome uncertain, given the constitutional provision that any base agreement *must* be ratified by the Philippines Senate, which can put it to a national referendum. Furthermore, even if some kind of new base agreement is reached, there are potential constraints on the freedom of U.S. activities from the Philippines that would change the U.S.-Soviet distribution of power in the region. For example, the draft constitution already contains a NWFZ provision for the Philippines.

Some Filipinos and Americans have sought to positively link the termination of U.S. base rights in the Philippines to negotiations for a reduction of the Soviet presence in Vietnam.[60] Certainly, the USSR in its desire to see U.S. strategic forces out of the Philippines has not discouraged such speculation. In his now celebrated July 28, 1986, Vladivostok speech outlining Soviet Asian policy, General Secretary Gorbachev stated: "In general, I would like to say that if the United States were to give up its military presence in the Philippines, let's say, we would not leave this step unanswered."[61] This has been diplomatically followed up in Manila, including a planned visit by senior Filipino defense officials to Vietnam in 1987.

Our discussion of South China Sea conflict zones demonstrates that although the roots of the contemporary problems of sovereignty and jurisdiction rest in the history of colonialism, the politics of resources, and the evolution of law, the issues are firmly embedded in broader political and strategic considerations. Whether the South China Sea becomes a zone of peace is not to be settled by Vietnamese and Filipino marines in the Spratlys or their analogues elsewhere, but by the security interests of the United States, the USSR, and the People's Republic of China in supporting or opposing them, and the capabilities they have to influence the outcome. It has been the great power contribution to the status quo that has helped to maintain it. The U.S. military facilities in the Philippines have been an important part of that contribution. The pattern will change if the status quo is terminated. In that event, Gorbachev's commitment to Asian peace and security will be tested.

Notes

1. This discussion of conflict zones in the South China Sea does not consider the problem of the territorialization of the Straits of Malacca. For this, see Michael Leifer, *Malacca, Singapore and Indonesia*, Vol. II, *International Straits of the World* (Alphen, the Netherlands: Sijthoff and Noordhoff, 1978); Yaacov Vertzberger, "The Malacca/Singapore Straits," *Asian Survey*, Vol. 22, No. 7 (July 1982), pp. 609–629.
2. Lee Yong Leng, *Southeast Asia: Essays in Political Geography* (Singapore: Singapore University Press, 1982), p. 112.
3. In general for energy resources in the South China Sea basin, see Selig S. Harrison, *China, Oil, and Asia: Conflict Ahead?* (New York: Columbia University Press, 1977); Corazon M. Siddayao, *The Offshore Petroleum Resources of Southeast Asia: Potential Conflict Situations and Related Economic Considerations* (Kuala Lumpur: Oxford University Press, 1978); Mark Valencia, "South China Sea: Present and Potential Coastal Area Resource Use Conflicts," *Ocean Management*, 5 (1979), pp. 1–38; and Kusuma Snitwongse and Sukhumbhand Paribatra, eds., *The Invisible Nexus: Energy and Asean's Security* (Singapore: Executive Publications, 1983). A short description of the petroleum factor is Mark Valencia's "Oil Under the Troubled Waters," *Far Eastern Economic Review*, March 15 1984, pp. 30–33.
4. *United Nations Convention of the Law of the Sea*, Article 123, "Cooperation of States bordering enclosed or semi-enclosed seas."
5. For an analysis from a regional vantagepoint of the South China Sea's geostrategic importance, see Lim Joo Jock, *Geo-Strategy and the South China Sea Basin: Regional Balance, Maritime Issues, Future Patterns* (Singapore: Singapore University Press, 1979).
6. Dr. Rocco M. Paone, "The New Chinese Navy," *American Intelligence Journal* (July 1984), pp. 5–15.
7. Bradley Hahn, "South-East Asia's Miniature Naval Arms Race," *Pacific Defence Reporter* (September 1985), pp. 21–24.
8. For general reference to Law of the Sea issues in Southeast Asia, see Lee Yong Leng, *Southeast Asia and the Law of the Sea: Some Preliminary Observations on the Political Geography of Southeast Asian Seas* (Singapore: Singapore University Press, rev. ed. 1980), and Phiphat Tangsubkul, *ASEAN and the Law of the Sea* (Singapore: Institute of Southeast Asian Studies, 1982). A concise overview is given by Phi-

phat Tangsubkul and Frances Lai Fung-wai, "The New Law of the Sea and Development in Southeast Asia," *Asian Survey*, Vol. 23, No. 7 (July 1983), pp. 858–878.
9. As quoted by David Jenkins, "Trouble Over Oil and Waters," *Far Eastern Economic Review*, 7 August 1981, p. 29.
10. A full exposition of China's historical claim to all four groups of the South Sea islands is given by Shih Ti-tsu, "The South Sea Islands Have Been China's Territory Since Ancient Times," *Kwangming Daily*, 25 November 1975, repeated by New China News Agency, 26 November 1975 (Foreign Broadcast Information Service, *Daily Report: People's Republic of China*, 28 November 1975), pp. E 1–8.
11. Hasjim Djalal, "Conflicting Territorial and Jurisdictional Claims in the South China Sea," *Indonesian Quarterly*, Vol. 7, No. 3 (July 1979), p. 42.
12. For the island disputes, in addition to the literature cited in note 2, see Phiphat Tangsubkul, "East Asia and the Regime of Islands," *South-East Asia Spectrum*, Vol. 4, No. 3 (April–June 1976), pp. 51–57; Hungdah China, "South China Sea Islands: Implications for Delimiting the Sea Bed and Future Shipping Routes," *China Quarterly*, 72 (December 1977), pp. 743–765; Martin H. Katchen, "The Spratly Islands and the Law of the Sea: 'Dangerous Ground' for Asian Peace," *Asian Survey*, Vol. 17, No. 12 (December 1977), pp. 1167–1181; Park Choon-ho, "The South China Sea Disputes: Who Owns the Islands and Natural Resource?", *Ocean Development and International Law*, Vol. 5, No. 1 (1978), pp. 27–59; Mark J. Valencia, "South China Sea: Present and Potential Coastal Area Resource Use Conflicts," *Ocean Management*, 5 (1979), pp. 1–38; Justus M. Van der Kroef, "Competing Claims in the South China Sea," *Lines of Communication and Security* (Proceedings of the National Defense University 1981 Pacific Symposium), pp. 25–50; and Michael Richardson, "Watch on the Spratlys," *Pacific Defence Reporter*, Vol. 6, No. 3 (September 1984).
13. Harrison, *China, Oil, and Asia*, p. 194.
14. A full explication of Vietnam's legal claims to sovereignty in the Paracels and the Spratlys is contained in Minh Nghi, "International Law and Vietnam's Sovereignty Over the Hoang Sa and Truong Sa Archipelagoes," *Vietnam Courier*, August 1982. See also Luu Van Loi, "Hoang Sa and Truong Sa: Vietnamese Territory," *Vietnam Courier* (September 1982), which directly refutes the Chinese position as laid out in the January 1982 Chinese Foreign Ministry's "White Paper" on the subject.
15. In effect, Manila took over the adventure of a Filipino entrepreneur who sought privately to colonize the northern Spratly's in the mid-1950s. Military occupation began in 1968. Presidential Decree

1596 in 1978 annexed the islands to Palawan Province, declaring them vital to the security and economic interests of the nations. Referencing history and discovery, including a claimed sighting by Magellan, it is stated that the islands are located on the Philippines' continental margin, ignoring the ocean trench between Palawan and the Spratlys.
16. *Quan Doi Nhan Dhan*, 26 April 1978, as reported in Foreign Broadcast Information Service, *Daily Report: Asia and the Pacific*, 28 April 1976, p. K 1.
17. Bernard Wideman, "Manila, Hanoi Beefing Up Forces on Disputed Isles," *Washington Post*, 10 June 1977.
18. Sheilah Ocampo-Kalfors, "Easing Towards Conflict," *Far Eastern Economic Review*, 28 April 1983, pp. 38–39.
19. Abby Tan, "Disputed Spratly Islands 'Vital to ASEAN's Defence,'" *Straits Times*, 15 March 1983.
20. Nancy Ching, "Chinese War Games in the Spratlys in 'Reply to Viets,'" *Straits Times*, 24 May 1983.
21. "KL Troops on Reef," *Straits Times*, 15 September 1983.
22. K. Das, "Perched on a Claim," *Far Eastern Economic Review*, 29 September 1983, pp. 40–41.
23. "Malaysia Rejects Viet Claim," *New Straits Times*, 11 June 1980.
24. Cheong Mei Sui, "Conflict of Claims Over Islands," *New Sunday Times*, 27 January 1980.
25. "Labuan to Resume 128-Year-Old Naval Role," *Straits Times*, 10 April 1984.
26. *Malaysian Digest*, 30 April 1984.
27. "The Labuan Issue," *Far Eastern Economic Review*, 4 September 1986, pp. 14–15.
28. For background and development of the Sabah dispute, see Chapter I, "Philippines Foreign Policy and the North Borneo Claim," in Bernard K. Gorden, *The Dimensions of Conflict in Southeast Asia* (Englewood Cliffs, N.J.: Prentice-Hall, 1966), pp. 9–41; Lela Garner Noble, *Philippine Policy Toward Sabah: A Claim to Independence* (Tucson, Ariz.: University of Arizona Press, 1977).
29. As late as May 1986, the regional press was reporting MNLF activities from a rear base on an island off Sandakan in northeast Sabah (*Straits Times*, 17 May 1986).
30. Only Indonesia has a true "continental margin" (shelf + slope + rise), and that is less than 200 nautical miles. For a succinct discussion of the geomorphology of the Continental Shelf problem in Southeast Asia, see Chapter III, "UNCLOS III and Continental Shelf Problems," in Lee Yong Leng, *Southeast Asia*, pp. 60–72.
31. An interesting Indonesian discussion of the politics of the new zones

is Asnani Usman, "Konflik Batas-Batas Teritorial di Kawasan Perairan Asia Timur," *Analisa*, Vol. 10, No. 2 (February 1981), pp. 125–150.
32. For the development of Indonesia's position from its original declaration of its archipelagic baselines in December 1957, see Mochtar Kusumaatmadja, "The Legal Regime of Archipelagoes, Problems and Issues," in Lewis Alexander, ed., *The Law of the Sea: Needs and Interests of Developing Countries* (Proceedings of the Seventh Annual Conference of the Law of the Sea Institute, University of Rhode Island, 1972), and Hasjim Djalal, "Indonesia and the New Extents of Coastal State Sovereignty and Jurisdiction at Sea," *Indonesian Quarterly*, Vol. 7, No. 1 (January 1979), pp. 80–93. The Philippine case is argued in Juan M. Arreglado, "Philippine Territorial Waters," *Decision Law Review*, 16 (1960), pp. 3–18, 83–101.
33. The Indonesian justification is part of its concept of the *wawasan nusantara*, which proclaims the inherent indivisibility of the total archipelago environment in both its physical and human manifestations, and ideologically underpins strategic thinking. See the discussion of *wawasan nusantara* by Donald E. Weatherbee, "Indonesia: A Waking Giant," in Rodney W. Jones and Steven A. Hildreth, eds., *Emerging Powers: Defense and Security in the Third World* (New York: Praeger, 1986), pp. 132–135.
34. Asnani Usman, "Masalah Penetapan Batas Zona Ekonomi Ekskulsif 200 Mil Indonesia," *Analisa*, Vol. 10, No. 8 (August 1981), pp. 712–733.
35. *Kompas*, 2 September 1982.
36. ASEAN state practices with respect to their continental shelves is discussed in Tangsubkul, *ASEAN and the Law of the Sea*, pp. 93–103.
37. "KL, Bangkok 'Nod' for Joint Authority's Functions," *The Sunday Times*, 25 October 1981.
38. The overlap comes as a result of the Philippines' claiming as territorial seas its "historic waters" as delimited by treaty. Pulau Miangas (Palmas Island), although outside the Philippines baselines, is situated in the southeast quadrant of the Treaty of Paris "treaty limits," and is considered by Manila to be the limit of its territorial waters. Miangas is at the same time the northern limit of Indonesian archipelagic baselines in East Indonesia. At the time of Spain's cession, the island was under Netherlands Indies administration. The issue between the United States as successor state to Spain and the Netherlands was put to arbitration, and in April 1928, the Hague Permanent Court of Arbitration decided in favor of the Netherlands.
39. It is Indonesia's firm policy to set median lines at points equidistant

from territorial baselines. Although the Philippines projects its jurisdictions from its own archipelagic baselines—not the treaty limits—to admit the Indonesian position would be to give up its "historical waters."

40. "Economic Zone 'Can Avoid Sea Rows in ASEAN,'" *Straits Times*, 6 September 1985.
41. "Declaration on the Baseline of Vietnam's Territorial Waters," *Vietnam Courier* (December 1982).
42. As reported by Xinhua, 28 November 1982. Foreign Broadcast Information Service, *Daily Report: China*, 29 November 1982; "Sea Frontier: Peking-Hanoi Issue," *New York Times*, 29 November 1982.
43. Van der Kroef, "Competing Claims in the South China Sea," *op. cit.*, p. 29.
44. Nayan Chanda, "All At Sea Over the Deeper Issue," *Far Eastern Economic Review*, 3 February 1978, p. 23.
45. *Indochina Chronology*, Vol. 1, No. 3 (July–September 1982), p. 3.
46. "Thais Reject Viet Claims on Territory," *Straits Times*, 23 November 1985.
47. Djalal, "Conflicting Territorial and Jurisdictional Claims," *op. cit.*, pp. 44–46.
48. For Indonesian capabilities and strategy in the South China Sea region, see Weatherbee, "Indonesia: A Waking Giant," *op. cit.*, pp. 156–158.
49. For the patterns of "regional resilience" and the Indonesian-Malaysian military connection, see Donald E. Weatherbee, "ASEAN: Patterns of National and Regional Resilience," in Young Whan Kihl and Lawrence E. Grinter, eds., *Asian-Pacific Security: Emerging Challenges and Responses* (Boulder, Colo.: Lynne Rienner Publishers, 1986), pp. 201–224.
50. *Straits Times*, 4 December 1985.
51. Manila Domestic Service, 15 March 1979, as reported in Foreign Broadcast Information Service, *Daily Report: Asia and the Pacific*, 17 March 1978, p. P 1.
52. "KL, Hanoi Agree on 'Friendly' Approach," *Straits Times*, 5 October 1983.
53. "No Let Up in Talks with Hanoi on Isles," *Straits Times*, 20 December 1983.
54. The text as given by Vientiane Radio, 7 October 1981 (as reported in Foreign Broadcast Information Service, *Daily Report: Asia and Pacific*, 14 October 1981, p. I 10), is printed as Document VIII in Donald E. Weatherbee, ed., *Southeast Asia Divided: The ASEAN-Indochina Crisis* (Boulder, Colo.: Westview Press, 1985), pp. 111–113.

55. "ASEAN to Be Nuclear Free," *New Straits Times*, 14 September 1984.
56. *Kompas*, 15 January 1985.
57. For a discussion of ZOPFAN and the maritime zones, see Mark J. Valencia, "Zopfan and Navigation Rights: Stormy Sea Ahead," *Far Eastern Economic Review*, 7 March 1985, pp. 38–39.
58. "Chinese Navy Build Up in South-East Asia," *Straits Times*, 24 November 1980.
59. Djalal, "Conflicting Territorial and Jurisdictional Claims," *op. cit.*, p. 42.
60. A most articulate exponent of this view before an audience of American and Filipino academics and officials was Ambassador Narciso G. Reyes, President of the Philippine Council for Foreign Relations, at the October 1986 Fletcher School Conference on "A New Road for the Philippines."
61. Text as given in *Current Digest of the Soviet Press*, Vol. 38, No. 30, 27 August 1986.

7. Thai-Vietnamese Rivalry in the Indochina Conflict*
William S. Turley

INDOCHINA has been a zone of almost continuous conflict for over forty years. Arguably the world's "most critical" threat to peace in the mid-1960s,[1] turbulence on the peninsula now involves only regional actors in direct conflict. On December 25, 1978, Vietnamese forces swept into Cambodia and evicted the Khmer Rouge regime of Pol Pot from Phnom Penh. Two months later, China launched a punitive attack across Vietnam's northern border. Since that time Vietnamese troops have continued to occupy Cambodia, while China, the United States, Thailand, and the Association of Southeast Asian Nations (ASEAN) have supported a coalition of Khmers fighting the Vietnamese in what has come to be known as the Third Indochina War.

This conflict is subordinate to rivalry among the great powers[2] in an important sense that one question makes obvious: What would happen if either of the principal regional contestants suddenly lost its great power support? It takes no analytical sophistication to predict that a sharp asymmetry of great power

---

*Some research for this paper was conducted while I was Visiting Professor at the American Studies Program, Chulalongkorn University, under auspices of the John F. Kennedy Foundation of Thailand and the Fulbright Program, 1982–84. During two trips to Hanoi in 1983 and 1984, my host was the International Relations Institute of the SRV Ministry of Foreign Affairs. I presented a revised version of the paper at Northern Illinois University on July 28, 1986, courtesy of the NIU Graduate School, Department of Political Science, and Center for Southeast Asian Studies. I should like to thank Professors Chai-anan Samudavanija, Donald K. Emmerson, Hans Indorf, Clark Neher, and Sheldon Simon for their helpful comments, and Professor Sukhumbhand Paribatra for providing me with copies of his papers on related topics in draft form.

involvement would cause a corresponding asymmetry of determination and capability among the local contestants. The conflict would end or at least abate in a form one side hitherto had judged unacceptable. But it is extraordinarily unlikely under current circumstances that any great power would make the necessary surrender of reputation, prestige, influence, and access. For abandoning clients, the Soviet Union would stand to lose its most important strategic gain since World War II, China its leverage in an area of vital security interest, and the United States its credibility as a treaty ally. Therefore, the regional contestants can count on the continuation of support they need to pursue their own objectives. Cloaked in and sustained by others' quarrels, regional actors contend for local advantage and pursue objectives that differ from those of their supporters. The concept of "proxy war" has little power to explain or predict in this situation.

The Third Indochina War can be understood in schematic form as a complex of conflicts involving four sets of contestants arrayed in concentric circles divided by one central cleavage. Each circle is related to its neighbor by ties that are analogous to those of patron and client. At the epicenter—for the present—are the People's Republic of Kampuchea (PRK) and the Coalition Government of Democratic Kampuchea (CDGK), the most abjectly dependent clients, patrons to no one. In the second circle are Thailand and Vietnam, the "front-line states" that, aside from the Khmer, have the most at stake and most directly support the Khmer. Thailand and Vietnam also enjoy support from both of the outer two circles. The circle representing the regional level is divided into the Indochina and ASEAN blocs, diplomatic cheering sections for the Vietnamese and the Thais, and regional coalitions aligned with great powers in the fourth circle.

Obviously, the various actors are involved in the conflict in very different forms and degrees, and this diagram glosses over these differences. But there is an overall arrangement of actors into four distinct arenas of conflict, each arena linked to another by ties of sponsorship and dependency. The central cleavage divides all the actors over certain lowest common denominators, but more significantly it divides the actors within each circle over issues of specific concern to them. Thus it divides the Khmer over the issue of legitimacy in Cambodia, Thailand and Vietnam over their respec-

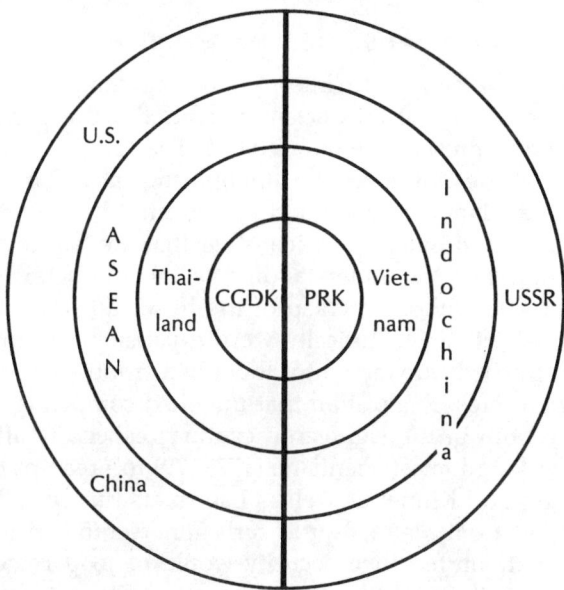

tive national security interests, ASEAN and Indochina over regional order, and the great powers over the global balance of power. Each concentric circle is an arena of conflict distinguishable from, though related to, the others.

This chapter contends that the key arena is that which pits Thailand against Vietnam. Without slighting the important role played by other powers with respect to Cambodia, it focuses on the larger contest for power between the two regional states. The rivalry of these two countries is not only intense and durable, it also determines the opportunities for others to become involved. Extending beyond the Cambodian question, it is likely to persist in some form regardless of how that conflict ends. Though in the present Vietnam directly threatens Thailand, in broader perspective neither country is an innocent victim of the other. The interaction of historical experience, mutual perceptions of intention, domestic politics, security strategies and capabilities have placed the two countries in a state of intractable confrontation that assumptions of rationality are inadequate to explain. In this context, a prudent U.S. diplomacy would be one that seeks to contain Thai-Viet rivalry and to avoid regional polarization.

I. Historical Experience

It is grossly oversimple to interpret the contemporary conflict only as an outbreak of an ancient rivalry. But the parallels between past and present are no accident. The terrain is the same, historical analogies influence perceptions and calculations in both Bangkok and Hanoi, and both the Thai and Vietnamese leaderships respond to images of each other that are based partly on historical learning. No understanding of the key actors' perceptions is possible without a grasp of the historical context.

Siam and Vietnam historically were expansionist, centralizing states that pushed outward into ill-defined frontiers and against crumbling empires. Realization that they had competing interests dawned on both in the eighteenth century, especially after Siam invaded the kingdom of Vientiane (1778–79) to preempt Burmese encirclement, and Khmer as well as Lao rulers turned to Vietnam for protection from Siam. By the early nineteenth century, both Siam and Vietnam felt their security would be jeopardized if the states between them were left free to cooperate with major enemies. Both sides practiced "territorial diplomacy," or expansion into marginal areas when strong, so that subsequent concessions when weak would leave their cores intact.[3] Both sought to exclude the influence of the other from Laos and Cambodia. And while the Vietnamese cast themselves in the role of cultural missionaries to the Lao and Khmer, the Siamese developed a sense of destiny to unite all peoples of Thai ethnicity. As of the mid-nineteenth century, Siam was ascendant, having absorbed all Lao territories (and the bulk of the Lao population) on the right bank of the Mekong, brought much of southern Laos under its direct administration, and made vassals of the remnant Khmer empire and all other Lao principalities plus the T'ai-inhabited Sip Song Chau Thai and Hua Phan Ha Than Ghok regions of northern Vietnam.

The French conquest of Indochina denied Siam its gains. But Siam's rulers, though resentful of France's domination of Laos and Cambodia, came to see French rule of Vietnam as a blessing in disguise. In 1930, King Rama VII wrote in a letter to his ministers that French rule was "a 'safeguard' for Siam. No matter how much we sympathize with the Vietnamese, when one thinks of the danger which might arise, one has to hope that the Vietnamese will not easily escape from the power of the French."[4] Wariness of Chinese influence on Vietnamese anticolonialism and abhorrence of radicalism in general made the Thai elite quite ambivalent about Vietnamese nationalism.

The French conquest also forced the Thai-Viet rivalry into suspension, and the two countries turned inward. In both, authoritarian regimes imposed change from above. But while successive generations of Vietnamese sought to overthrow the colonial government, Siamese monarchs sought to "modernize" the kingdom without undermining royal absolutism and elite privilege. Preoccupied with attaining power. Vietnamese revolutionaries strove to organize broad-based mass support; holding a monopoly of power, the Thai elite circumscribed politics to the court and high-ranking bureaucrats. The Vietnamese Communists ultimately were to build a "people's army" and defeat the French, but the Thai Army ceased to wage war for a century as it became an instrument of internal royalist consolidation, not external defense.[5] These differences laid bases not only for sharply different political-economic systems but also for quite different security orientations on the part of modern elites in the two countries. To oversimplify, Vietnam's Communist leaders were to feel that the revolution's rootedness at home made them invincible on their own territory, whereas the Thai military and bureaucratic elites that ended royal absolutism lacked comparable links to society and would look abroad for help to suppress internal threats.

Another crucial dissimilarity arose in connection with the place each occupied in the other's consciousness. While the Vietnamese became absorbed in resistance to the West, Thailand played host to two waves of refugees from Vietnam. The first wave involved a flight of Catholics during the religious persecution of the 1830s who settled in Bangkok and were largely assimilated. The second wave involved some 50,000 people who left Indochina during the early stages of the war with France. The latter group has not been assimilated and the Thais continue to regard it warily because of its potential ties to a foreign power, its sinitic culture, and perceived economic dynamism. A significant result is that while the Thais exist only on the horizons of Vietnamese policymakers' perceptions, the Vietnamese seem a domestic as well as external threat to a broad spectrum of Thais.

For external security during the colonial period, Vietnam suffered France's "protection" while Thailand was a buffer between French and British possessions. Self-defense was moot for Vietnam, unnecessary for Thailand. The relative passivity of Thai diplomacy toward the neighboring colonial regimes, however, was only an expedient response to the power imbalance. As French power waned during World War II, the Thais, under the right-wing chauvinist Field Marshal Phibunsongkhram and with Japanese support, renewed claims to territory in Laos and Cambodia. In 1946–47, the left-leaning Pridi Phanomyong permitted Viet Minh and Lao revolutionaries to take refuge and purchase weapons in Thailand. Phibun briefly continued this policy after returning to power, since the right and left shared a desire to loosen the French grip.[6] Thus what one Thai scholar has called the "enduring logic" of excluding other powers from the trans-Mekong Basin reasserted itself as the colonial regime weakened.[7]

But this logic applied to the Vietnamese as well. Only by attacking the French—and promoting revolution—in Laos and Cambodia could the Communists prevent France from concentrating resources in all three countries against them. Strategic security requirements made both the Thais and Vietnamese feel deeply threatened by unfriendly or "neutral" regimes, political instability, and the presence of foreign powers in Laos and Cambodia.

When it appeared to the Thais that the Vietnamese Communists might supplant the French in all of Indochina, Thailand in 1950 signed a military assistance pact with the United States. Subsequently, in the Second Indochina War, Thailand cooperated with the United States in an attempt to prevent the reunification of Vietnam. The Thais sent twenty-five battalions of combat troops secretly into Laos, participated in the subversion of Laos' neutrality, dispatched a contingent of troops to South Vietnam, and permitted American planes to bombard North Vietnam from bases in Thailand. The Vietnamese Communists responded by supporting a Communist insurgency in Thailand which also enjoyed Chinese support. Though in the beginning they may have wished to see the Thai government fall, their relations with China turned sour and they may have sought, as they claim, only to oppose the American use of Thai territory. Meanwhile, of course, they sponsored revolutions in Laos and Cambodia that laid a basis for Vietnamese predominance in all of Indochina.

One of the legacies of this history for the Thais has been the survival of pan-Thai elements in Thai perception, if not aspiration. Dr. Thanat Khoman, architect of contemporary Thailand's foreign policy, in 1976 reminded the "new regimes in Indochina" that any claim by Laos to territory in northeast Thailand "would go against ethnic and historic facts as there are only five known races in Southeast Asia: the Thai, Mon, Malay, Burman and Khmer. There is no such thing as a Lao or Laotian race which is but a branch of the Thai race whose offshoots may be found in North Vietnam, Shan State, Assam, Laos and, of course, Thailand."[8] As recently as 1985, a Thai Foreign Ministry White Book on Laos maintained that the Thai and Lao peoples had failed "to unite into one single kingdom" only because the process of integration was interrupted by "Western colonial powers."[9] Bangkok's diplomacy toward Laos still tends to be alternately condescending and overconfident in the attractive power of a common culture. No revanchists, Thai leaders have clung to the assumption that cultural affinities are a strategic asset. They are aware that peoples of Thai ethnicity comprise mainland Southeast Asia's largest "nation," and while harboring no design to unite this nation, they assume this fact gives them potential leverage in relations with their neighbors. A second legacy has been reliance on peripheral "buffers" for external security. These "buffers" may be neutral states, or they may be insurgencies aimed against potentially threatening neighbors. It was the latter that Supreme Commander General Saiyud Kerdphol had in mind when he said in 1982 that Thailand had not lost its historic buffers in Laos and Cambodia because "fighting is going on. . . . Thailand's 'buffer' in that sense has not disappeared."[10] A third legacy is the perception of Vietnam as the successor to France in unifying Indochina. The French first accomplished the unification of Indochina, then withdrew in circumstances that left the Communists poised, it seemed to the Thais, to fill France's shoes. It makes no difference in this view whether the Vietnamese use "federation," a "special relationship," or alliance to unify Indochina, since it is Vietnamese military access to Laos and Cambodia that constitutes the threat.

As for the Vietnamese Communists, the principal legacy is the "lesson" that the nation's security and unity absolutely require regimes in Vientiane and Phnom Penh that are stable and aligned with Hanoi. Successive wars with great powers that utilized Laos and Cambodia to attack the Communists' only secure north-south link, and Chinese support of anti-Vietnamese forces in Laos and Cambodia, have made this lesson a "law of history" for the Vietnamese.[11] Second, as regards Thailand, Bangkok's alliance with the United States, facilitation of U.S. military strategy in the last war, and subsequent cooperation with China in the Third Indochina War have reinforced the Vietnamese propensity to perceive the Thais as chronically dependent for their security on great powers hostile to Vietnam. In Vietnamese perception, Thailand's dependency not only provides access to the region for great power adversaries; it also provides cover and encouragement for the Thais to strive for the restoration of their hegemony in Laos and Cambodia as well. Though Thai military elites supported American strategy in the 1950s, 1960s, and 1970s for purposes they perceived as defensive, in Vietnamese perception the effect was offensive and threatens to be so again. It is in such legacies that we can find the bases of mutual suspicions that make for durable tensions between these two countries today.

II. Perceptions of Intention

Mutual suspicions sustained by incompatible historical legacies exacerbate a classical example of the security dilemma in Thai-Viet relations. Although the two states share an interest in peace, development, regional stability, and avoidance of intrusion by great powers, the structure of the situation as they see it prevents them from bringing about the mutually desired outcome. The intensity of antagonism may wax and wane—the Second Indochina War was a high point—but each side's reading of the geopolitical realities remains relatively fixed. Each side takes steps it sincerely proclaims are defensive that the other views as menacing. Each has an image of the other's intentions that causes it to react in ways that perpetuate the impasse.[12]

For the Thais, who view the Mekong as a line of unity not division,[13] it is not Vietnam's military strength so much as the deployment of its forces in the Mekong Basin that is threatening. According to General Chaovalit Yongchaiyuth, currently army commander-in-chief, a strong Vietnam benefits Thailand by distracting China and is not likely to mount a "serious invasion," but the distribution of Vietnamese forces on opposite ends of the Indochina "strip" is an intrinsically threatening posture. Even if the Vietnamese have no intention to invade, in General Chaovalit's view their military emplacement positions them to intimidate.[14] Other Thai strategists who are more alarmed believe Vietnam has designs on Thailand's fourteen northeastern provinces, and regard the construction of road links from Vietnam to points along the Mekong and from Vientiane into Sayaboury Province, a salient of Laos that protrudes into Thailand, as evidence of this intention. "It is suspicious," one "senior military analyst" has said, "that two Lao divisions, one Vietnamese division, and two armored regiments are stationed there."[15]

Although some Thai leaders perceive benefits for Thailand in the Indochina conflict,[16] virtually all agree that the Vietnamese emplacement is an intolerable long-term threat. The Thais further perceive that Soviet support of Vietnam tilts the balance of airpower against them,[17] buys a permanent Soviet naval presence at Cam Ranh Bay,[18] and sustains Vietnam's determination and capability to dominate Indochina. They appear unanimous in believing that Hanoi's diplomacy aims to manipulate divisions within ASEAN for the purpose of isolating Thailand from its ASEAN partners. These perceptions provide the Thais with rationale to obtain great power guarantees, including if necessary from China, and to consider seriously any measure that might restore their "buffers."

The Vietnamese for their part do not regard Thailand as a threat in itself. Neither Thai military power nor Thai cultural and economic appeal to ethnolinguistic cousins—at which the Vietnamese scoff[19]—disturbs the Vietnamese. The Vietnamese also express unconcern that the wealth gap might grow between themselves and the Thais, since in their view Thailand's capitalist development causes income inequalities, class frictions, and political instability that dissipate the kingdom's energies.[20] In Hanoi's view the Thai threat consists of Bangkok's external relations, principally with China, secondarily with the United States. Although

Vietnamese officials offer different assessments, at high levels they do seem to subscribe to the line they purvey in propaganda: that China has acquired (or is acquiring) inordinate influence within the Thai political system through the manipulation of the Sino-Thai business community, appeal to Sino-Thais now rising to responsible positions in the bureaucracy, and cooperation with the Royal Thai Army in a wide range of activities (e.g., exchange of intelligence and visits, co-production of arms, Chinese advice on popular defense, recruitment and training of Lao insurgents). Foreign Minister Nguyen Co Thach has said, "As far as we are concerned, there is no argument between us and certain individual ASEAN states. . . . In a way, we wouldn't even be in conflict with Thailand, either, if China did not use it against the states of Indochina."[21] The Vietnamese nightmare is that, having kicked the Chinese out of Cambodia along with the Khmer Rouge, they may in the future find the Chinese ensconced in Thailand, presenting Vietnam once again with the prospect of a two-front war. Cognizant of the Thai elite's heavy dependence in the past on foreign patronage, the Vietnamese speculate that Bangkok is prepared to accept a measure of Chinese direction over its foreign policy as the price of mortally weakening Vietnam (as in the last war, in cooperation with the United States), attaining territorial objectives in Laos and Cambodia (as in World War II, with Japanese support), and splitting the Indochina bloc. Although Hanoi's allies in Phnom Penh are often the most explicit, the line is essentially the same in both capitals: Thailand has "expansionist and hegemonist designs" and is attempting "to disrupt the solidarity of the three Indochinese countries."[22] Or, as Phnom Penh's Deputy Foreign Minister Kong Korm replied to the question, What do you think the Thais really want?, "Thailand follows China's policies. The Thais have not forgotten that they had to cede Battambang, Siem Reap and Sisophon provinces to the French, and they think that by helping China to support Pol Pot they can benefit. They entertain the fantasy that by supporting reactionary forces they may gain territory in lost territories."[23] Vietnamese officials, queried along the same lines, are less emphatic but often respond quizzically, "Why do the Thais always cause trouble?" The imputed intentions may seem fantastic, but they are no more fantastic than Thai speculation about

Vietnamese designs to absorb Thailand's fourteen Lao-inhabited northeastern provinces into an "Indochina Federation." The Vietnamese, who consider themselves much stronger than the Thais, find it difficult to comprehend the apparent risk-taking of Thai policy without assuming that the Thais seek proportionate gains, which could only come at Vietnam's (and Laos and Cambodia's) expense. In perception, it is a zero-sum game.

The suspicions between Thailand and Vietnam would be rich terrain for students of mirror-imaging. Each denies it is the rival of the other. Each perceives the other as growing dependent on a great power for attainment of aggressive aims as well as for legitimate defense. Each claims the other seeks to split it from its regional allies. And each has responded by tightening relations with great powers, obtaining new weapons, and refusing accommodation, thus confirming the other's initial suspicion. The Vietnamese have attempted to allay Thai fears verbally and by what they claim is respect for the border (despite some possible warning incursions), but only a realignment of Hanoi's foreign relations and abandonment of military deployments in Laos and Cambodia would be credible in Bangkok. Likewise the Thais maintain that the "front line" is *not* between Vietnam and Thailand, rather it is between Vietnam and China, but the Vietnamese remain unconvinced so long as Bangkok cooperates with Beijing in supporting the Khmer resistance. Thailand requires that Vietnam withdraw its forces from Cambodia, give non-Communists a share of power in Phnom Penh, and face the risk of instability on its western flank; Vietnam requires that Thailand cease supporting Khmer forces, accept Hanoi's fait accompli, and face the risk of angering China. Neither feels it can placate the other on terms it can afford to accept.

III. Domestic Politics

Thailand

At the last war's end in 1975, domestic factors caused both Thailand and Vietnam to grope toward coexistence. Vietnam's need for peace was plain enough, while the forces at work in Thailand were more complicated. Although Bangkok already had begun disengaging from U.S. policy, the student uprising that overthrew military rule in 1973 left a very fragile civilian leadership to cope with the implications of American withdrawal. To counterbalance the potential threat from Vietnam's reunification, Bangkok improved relations with Beijing, and in October 1975 with Chinese help it became the first non-Communist country to establish diplomatic relations with Pol Pot's Cambodia. More than their autocratic predecessors, the elected civilians also felt constrained to mobilize domestic support for such key foreign policy objectives as a strengthened ASEAN and favorable investment climate.[24] These were the bases of the Kukrit Pramot government's advocacy of "equidistance" in relations with great powers. Ultra-rightist military figures seized power in 1976, but their overzealous anticommunism provoked diverse groups spearheaded by "Young Turk" officers into uniting behind the moderate General Kriangsak Chomanand.[25] Kriangsak revived the flexible, pragmatic approach and made friendly overtures to Laos in 1978. But Vietnam's invasion of Cambodia brought the rapprochement to a halt. With Vietnamese troops on their border, the Thais agreed to let China use Thai territory to resupply the Khmer Rouge. Since that time domestic processes have hardened the antagonism and made compromise more difficult.

In Thailand, the hardening began with the "psychological trauma" that resulted from the "prospect of sharing a *de facto* common border with Vietnam running some 1,250 miles from Laos to Kampuchea."[26] That trauma helped to consolidate a very high degree of elite consensus not to tolerate Vietnam's presence in Cambodia. In a 1982 survey of elite figures, 97.3 percent of respondents cited Vietnam as a threat (the highest percentage of positive responses), 74.3 percent identified the form of Vietnam's threat as direct military aggression, and only 1.1 percent indicated willingness to accommodate Vietnamese domination of Cambodia as a way to deal with the issue.[27] Elite consensus has helped to keep the issue out of partisan politics, and seldom has anyone voiced concern over established policy or the military and government's handling of it.[28] Combined with rising confidence that a long conflict harms Vietnam more than Thailand,[29] the consensus effectively restricts policy discussion to the question of means.

The elections and party realignment that took place in July

1986 did little to change this. Foreign Minister Air Chief Marshal Sitthi Sawetsila retained his post and announced his intention to continue existing policies concerning Indochina, to make no new proposals, and to leave negotiations on a Vietnamese pullout from Cambodia to the initiative of the CDGK.[30] In an interview, Sitthi declared that "Thailand will not allow its territory to be used to destabilize or undermine the governments of neighboring countries,"[31] but the Foreign Ministry "clarified" Sitthi's remark by stating that the Heng Samrin regime "does not represent Cambodia," and that Thailand along with ASEAN and the UN continued to "consider the government of Prince Sihanouk to be the legitimate government of Cambodia."[32] Sitthi did indicate the new government would be "more independent" in certain areas of foreign policy, but this appeared to apply primarily to economic relations with major powers.[33] The government's perception of need to improve the business climate did not imply concessions over Cambodia, though Bangkok did move to relax diplomatic relations and trade restrictions with Laos[34] and seems likely to continue tolerating a very small trade with Vietnam itself.[35]

Furthermore, Thailand remains in many respects a bureaucratic polity, where internal factors long have had predominant influence on foreign policy.[36] Basic decisions are made within the bureaucracy and reflect the relative power of competing cliques. In this competition the military prevailed for decades, sometimes to the point of conducting foreign policy without the knowledge, much less the approval, of the Foreign Ministry.[37] However, the military has never completely recovered politically from the events of 1973–76, due to the drying up of patronage resources previously drawn from U.S. military assistance, the growing assertiveness of an expanding middle class, and political fragmentation within the military itself. Also, by successfully suppressing the Thai Communist insurgency, the military has deprived itself of the mission it used for years to justify the supremacy of security in national priorities. Against this background the Indochina conflict provides military leaders with rationale to demand, if not a garrison state, then one in which they would continue to have significant institutionalized participation in policymaking. Army commanders who sought in April 1983 to amend the constitution in order to allow uniformed officers to hold political office justified their demand by citing the external threat. Of course that threat also supplies the military with argument for increased defense spending, suppression of internal dissent, formation of paramilitary organizations, and other manifestations of the "national security state."[38] It may not be going too far to say that some Thai military leaders have a domestic political interest in protracting the conflict on Thailand's borders.

Though the shift to an external security mission has provided an impetus for military professionalism, the tendency to protect personal and clique interests in the political realm remains strong. It is instructive in this regard to consider the case of the Democratic Soldiers, an army group that virtually alone opposed the government's Cambodia policy in the early stages of Vietnam's occupation. The group prophesied that China would exploit conflict between Thailand and Vietnam to win power for the CPT, and recommended that Thailand remain neutral.[39] Suspicion of China and obsession with insurgency were pervasive in the army, especially among officers like most of the Democratic Soldiers who had long career involvements in the fight against the CPT. However, the Democratic Soldiers disbanded in late 1981, and a senior officer with close ties to them, General Chaovalit Yongchaiyuth, utilized money obtained from control of the border black market to help found one of the anti-Vietnam resistance groups, Son Sann's Kampuchea People's National Liberation Front.[40] Money from the border also financed the abortive "Young Turks" coup in April 1981. In 1983, an estimated 5 million *baht* (U.S. $217,000) a day flowed from the border into the Army Operations Center, from which General Chaovalit disbursed it to support the Khmer resistance and other political projects.[41] General Chaovalit became army commander-in-chief in May 1986 and is widely believed to aspire to succeed Prem Tinsulanond as prime minister. Just how this background relates to General Chaovalit's promotion and views is unclear, but it is hardly consistent with neutrality toward the Cambodian conflict or indifference to the exploitation of that conflict for domestic purposes.

In a similar vein, officers assigned to Task Force 80, the Thai Army's special border command, reportedly have refused transfer to positions that offered better chances of promotion because service on the border was so much more financially rewarding. Civilians, too, have benefited from the border's black market trade and local spending by relief organizations.[42] An economically depressed area from 1975 to 1979, the border region recovered thanks to the influx of money and people. The nexus of money and power on the border is extremely murky, but it almost certainly works in ways that sustain the commitment to current policy, and it could provide military interests with a motive to veto meaningful initiatives.

Vietnam

As for Vietnam, the closed domestic political process is much more subject to speculation, but it appears to have had the same hardening effect as in Thailand. Foreign policy was the subject of hot debate at the Fourth Party Congress in 1976, which resolved to oppose "great power chauvinism." Following the crises of 1978–79, a purge removed individuals who preferred accommodation with China and consolidated support for the intervention in Cambodia. Whatever foreign policy factionalism existed came to an evident end. Moreover, the People's Army has benefited enormously from the close ties with the Soviet Union that have been forged since Moscow and Hanoi signed a treaty of friendship and cooperation in November 1978. Without Soviet assistance, the Vietnamese military would be unable to fight for long in Cambodia. Even more important, it would be unable to modernize its technology and improve its conventional war-fighting capability—crucial objectives of officers who would like to put "people's war" behind them.[43] So long as the Soviets seek to "contain" China, the Vietnamese military must cooperate in that objective to justify Soviet assistance, and it may well look upon confrontation as an opportunity to serve its institutional interests. The army also has come to see its reputation for invincibility staked on a military victory in Cambodia. Officers who bask in the enormous prestige of the People's Army would not likely support diplomatic initiatives that could be construed as retreat. Conceivably new leaders committed to economic reform and development, such as Nguyen Van Linh, may favor concessions (though there is no hard evidence of this to date), but they may also need to prove their militancy to retain power. As in Thailand, domestic factors almost certainly limit how far Hanoi can move toward compromise without significant internal political change.

IV. Strategies and Capabilities

Thai and Vietnamese strategies exacerbate the suspicions that underlie the rivalry. Both parties perceive each other to be serving in some degree as the instrument of a great power that is the "real enemy." Each also believes the other's determination depends heavily on continued great power support. Each party perceives its adversary's efforts to cope with that larger threat as threatening to itself. Thus the Thais feel compelled to protect themselves from Vietnam's preparations to fight China and the Vietnamese to protect themselves from Thailand's countermeasures that facilitate China's influence in Southeast Asia. However, without leverage on their opponents' great power allies, the two regional actors can maneuver effectively only against each other.

Thailand's strategy, which is more evident in action than public statement, is to forestall Vietnam's establishment of absolute dominion over Laos and Cambodia while the kingdom continues to consolidate its internal security, political stability, and economic development, thus altering the balance in Thailand's favor over the long term. Unable to do this alone, the Thais have joined with China, the United States, Japan, and ASEAN to isolate Vietnam diplomatically and economically, and with China to support Khmer and Lao insurgents. They also have welcomed China's security guarantee while striving to rekindle American interest. Though the Thais say they have no desire to weaken Vietnam permanently because a strong Vietnam will help to distract China in future,[44] it serves Thai interests for a time to have Vietnam bogged down in a costly confrontation while Thailand continues to grow. (Thailand's GNP already is about three times that of Vietnam.) The economic disparity will help Thailand to offset Vietnam's military superiority in decades to come. Meanwhile, the Thais are pleased to note that confrontation with Vietnam has helped them to obtain American reassurance, converted China from an enemy into a friend, helped to liquidate the Thai Communist insurgency, and strengthened cooperation among the six ASEAN nations.[45] As the "front-line state," Thailand also has enjoyed status and influence within ASEAN that it never had before. Although the strategy could backfire by alienating Indonesia, up to the present it has brought Bangkok important benefits at little cost. Even if this strategy does not succeed in loosening Vietnam's hold on Laos and Cambodia, it prepares Thailand to cope with the threat of a unified, Soviet-supported Indochina.

This is not to imply that Thai strategists are resigned to Vietnam's eventual success. On the contrary, many Thais are prepared to pay a higher price than they have paid up to now to prevent any form of Indochinese unification under Vietnamese hegemony. Some military figures in particular seriously contemplate more vigorous measures. They envision a Khmer insurgency capable, not of ejecting the Vietnamese altogether, but of holding a "liberated zone" that would suffice as Thailand's buffer, perhaps in the form of a divided state. Such an insurgency would require unstinting Thai, Chinese, and American support. It also might require the intervention of Thai armed forces, which Thai officers discuss openly. General Pichit Kullavanijaya, commander of the First Army Region, told this writer in 1984 that only three outcomes in Cambodia were conceivable—complete control by Vietnam, ejection of the Vietnamese by the Khmer resistance, and division into "two zones or countries"—and of these only the third was both possible and acceptable. Asked if that ruled out negotiated compromise as advocated by ASEAN and the Thai Foreign Ministry, the general replied, "No, a negotiated compromise is possible if it consists of partition."[46]

For the foreseeable future, scenarios of intervention and partition lie in the realm of fantasy. However, the discussion of them suggests how deeply some Thais feel their security depends on territory from which Vietnam's political influence as well as military presence are entirely excluded. Developments that lent plausibility to these scenarios—e.g., increased American involvement, renewed vigor in the Khmer resistance, a slackening of Soviet support for Vietnam—could gain them a wider hearing. Needless to say, from the Vietnamese point of view these currents in Thai strategic thought seem plainly hostile.

Vietnam's strategy is to remove forever the possibility of threat from across the Annamite Cordillera by creating a bloc of three Indochinese states. This strategy requires Hanoi to tutor regimes like its own in Laos and Cambodia and to promote political, military, economic, and cultural cooperation within Indochina.[47] Hanoi may grant Laos and Cambodia greater latitude for independent policymaking as time goes by, if only to cut its costs, but never so much as to jeopardize the ideological unity and security interdependence of the three states.

In Cambodia, Vietnam's dilemma has been to create a Khmer regime and army without so dominating them that they never gain legitimacy and self-confidence. Strategy since 1983 therefore has been to take the "calculated risk," according to Foreign Minister Thach, of gradually withdrawing Vietnamese forces. By making the population and Phnom Penh army more responsible for their own defense, Hanoi has hoped to wean the Khmer from dependence on Vietnam and to galvanize support for the Heng Samrin regime.[48] The unprecedented attacks on border encampments of the Khmer resistance during the 1984–85 dry season were intended to provide Phnom Penh with time to consolidate the "inland front" around the Great Lake so that this strategy could be carried out.[49] The objective is to turn the fighting over to the Khmer, withdraw the bulk of Vietnamese forces by the unilaterally stated goal of 1990, and hand the world a fait accompli. The strategy assumes that international opposition will fade away as China seeks to improve relations with the Soviet Union, Thailand finds itself at odds with ASEAN partners growing anxious to avert great power intrusion, and the issue slips off the global agenda or is resolved by negotiations that save face for ASEAN but meet Hanoi's terms.

Should this scenario not come to pass and "some countries in the region, especially Thailand, allow their territory to be used by foreign countries against a third country," the Vietnamese have made clear, they will delay their military withdrawal.[50] Thus Hanoi has made withdrawal contingent, in effect, on Thai acceptance of Vietnam's predominance in Cambodia. That certainly casts doubt on the likelihood of Vietnamese withdrawal by 1990.

The Vietnamese also could raise the ante for the Thais by resuming support for a Communist insurgency in Thailand's northeast. Though such an insurgency now would have to be resurrected from virtual extinction, the Thais anticipate the attempt. According to Thai intelligence, Hanoi's efforts to entice the CPT away from its pro-China alignment in 1976–78 were unsuccessful,[51] but a small group of former CPT members and leftist students came together in 1979 to form a new party, the "Phak Mai," under Lao-Vietnamese auspices.[52] Subsequently, the Phak Mai and related groups are said to have recruited upwards of 500 members and to have sent small armed propaganda teams from Laos into Thailand's northeast to contact relatives and conduct reconnaissance, with Laotian and Vietnamese support.[53] If these reports are true—and Hanoi's previous support for the CPT makes them plausible—Vietnam may be preparing an insurgency option as bargaining leverage against the Thais. Given Bangkok's extreme sensitivity to instability in the Lao-inhabited northeast, even the appearance of tentatively preparing to support an insurgency has threat value for Hanoi.[54] It is not inconceivable that Vietnam could hold Thailand's hard-won internal security hostage to Bangkok's ending cooperation with China in Cambodia. Although the credibility of this threat recedes as Bangkok gains control of its hinterland, it would grow if economic and political problems undermined Thailand's stability. In February 1987 the Thai government geared up for a broad political offensive against remaining CPT appeals.[55]

Down to the present, a major constraint on both countries has been concern to avoid widening the conflict, incurring even greater risks. Another constraint has been limited capabilities. With 1.2 million men under arms, up to 170,000 men in Cambodia,[56] and forward supply depots near the Thai border, Vietnam's gross military strength is overwhelmingly superior to that of Thailand. But over half of Vietnam's forces are deployed to defend against China, and even those in Cambodia would not be able to sustain a large-scale attack for long outside their current perimeter. The Thais for their part, with 240,000 total armed forces including a 160,000-man infantry, are inferior to the Vietnamese in "every combat element . . . effectiveness and experience."[57] They are hardly likely to mount "offensive-defensive" strikes to preempt Vietnamese incursions as top officers threatened in March 1985. This situation is not likely to change soon. The Vietnamese have no ability to expand and upgrade their military without still greater Soviet support, while the Thais have been forced by falling commodity prices and currency devaluations to curb defense spending and manpower.[58] Neither party has sufficient resources to coerce the other or to break the stalemate without a very sharp increase of support from a superpower ally, which is not forthcoming. The anticipation that the buffer of incapability someday will cease to exist, however, already has caused the first stirrings of an arms race. Since 1979, Thailand has maintained the highest level of military expenditure as a percentage of the government budget of any ASEAN member, and it has purchased twelve F-16A/B aircraft specifically to help offset Vietnam's four-to-one advantage in fighter-interceptors.[59] Vietnam meanwhile has concentrated its procurement on such weapons as the MI-24 Hind gunship for use in counterinsurgency and naval craft for patrolling the South China Sea. The Vietnamese military also has sought to obtain MiG-23s to maintain air superiority on the Chinese border.[60] Though Hanoi has acquired or is attempting to acquire weapons with Cambodia and China in mind, the buildup provokes a counterresponse from Thailand, which before long may merit its own share of Vietnamese attention.

V. Conclusion: The American Factor

The foregoing discussion has attempted to show that Thailand and Vietnam are locked in rivalry by more than the rational calculation of their national self-interests. The fundamental assumptions on which Bangkok and Hanoi base their external policies, buttressed by perceptions and political processes in both capitals, scarcely permit a rapprochement that would leave one side with a monopoly of access to the countries between them. While Vietnam's drive to establish such a monopoly has been obvious, Thailand's determination to prevent it must not be underestimated. What has been characterized as a "stable war" in Cambodia because the great powers lack compelling reason to end it[61] is also subordinate to a contest of regional powers. In the absence of arrangements imposed and enforced by the great powers acting in concert—a presently inconceivable possibility—the long-range prospect for Indochina is more or less perpetual conflict, and for mainland Southeast Asia a similarly protracted state of tension.

Does that prospect require any change in the American approach toward the region? Since 1979, American policy toward Southeast Asia has responded almost solely to the growing Soviet role. The threat posed by Soviet use of naval facilities at Cam Ranh Bay and support of Vietnam heightened American concern for the continued use of bases in the Philippines, the alignments of ASEAN states, and the preservation of Sino-U.S. cooperation against the Soviet Union. Accordingly, the United States expanded its naval and air capabilities in the Pacific, encouraged Japan to rearm, and increased arms sales to ASEAN. While these measures provided ASEAN a welcome alternative to the choice between Soviet or Chinese influence, the "de facto joint Sino-U.S. security guarantee to Thailand" and U.S. support of Beijing's intransigent position on the Cambodian issue made some ASEAN members uncomfortable.[62] Indonesia in particular has felt that Sino-U.S. support of Thailand has stymied efforts to move beyond the Cambodian conflict, strike balanced relations with all three great powers, and implement the idea of ZOPFAN (Zone of Peace, Freedom, and Neutrality). There is also concern that Sino-U.S. cooperation facilitates the expansion of Chinese influence in the region.

The problem for the United States has been to reconcile its strategic relationship with China and commitments to Thailand with its interests in preserving ASEAN unity and strengthening relations with ASEAN's largest member. The effort to effect that reconciliation is likely to grow more difficult as time goes by. For, in the first place, U.S. interests in Southeast Asia are bound to clash eventually with those of China as the latter strives to assume principal responsibility for regional peacekeeping. Although China will face serious constraints due to its own weakness and the suspicion it arouses in the region for a long time to come, American passivity meanwhile abdicates the initiative to China, to the consternation of every ASEAN member. Second, a prolonged conflict in Cambodia in circumstances of minimal American involvement tends to push Thailand closer to China. This is not desirable from the standpoint of Thai-U.S. relations, the unity of ASEAN, or the prospects for peace on the peninsula. And third, current measures to bring about Vietnamese withdrawal and coalition government in Cambodia are inadequate. It is Vietnam's proposals, not ASEAN's, that are being implemented. Conceivably this could be changed by major direct American involvement, but in the certain absence of that prospect Vietnam's fait accompli is the likely outcome. Protracted yet limited efforts to avert that outcome merely deepen Vietnam's dependence on the Soviet Union, causing some ASEAN states to consider accommodation with Hanoi less risky than the extension of Sino-Soviet rivalry into the region. Thus American attempts to sustain ASEAN's determination from behind the scenes help to produce effects they are intended to avoid.

Since the United States has good reasons to avoid deep involvement in a leadership role, it must consider the alternative of helping the region to adjust to a new status quo. That alternative does not require the United States to "reward" Vietnam in any way, but it does require it to help Thailand, through security assistance and economic cooperation, to develop a stronger sense of self-sufficiency in defense without "buffers" or close ties to China. Over the long run American interests will be served best not by any particular outcome in Cambodia but by a Southeast Asia comprised of stable regimes, economies more advanced than China's, and nations at peace. Such a region would be the one envisioned by ASEAN statesmen in which rules of order were made by Southeast Asians themselves. It is a vision out of reach so long as Thailand and Vietnam contend for influence in the countries between them.

Notes

1. Feliks Gross, *World Politics and Tension Areas* (New York: New York University Press, 1966), pp. 192–193.
2. The term "great power" is preferred here because China is an essential actor in the regional context yet clearly is not a superpower.

3. See Robert L. Solomon, "Boundary Concepts and Practices in Southeast Asia," *World Politics*, 23 (October 1970), pp. 12–13.
4. Quoted in Benjamin A. Batson, *The End of the Absolute Monarchy in Siam* (Singapore: Oxford University Press, 1984), p. 117.
5. On the Thai military, see Benedict R. O'G. Anderson, "Studies of the Thai State: The State of Thai Studies," in Eliezer B. Ayal, ed., *The Study of Thailand: Analysis of Knowledge, Approaches, and Prospects in Anthropology, Art History, Economics, History, and Political Science* (Athens, Ohio: Ohio University Center for International Studies, Southeast Asia Series No. 54, 1978), pp. 200–205.
6. Evelyn Colbert, *Southeast Asia in International Politics, 1941–1956* (Ithaca, N.Y.: Cornell University Press, 1977), pp. 81, 100.
7. Sukhumbhand Paribatra, "Thailand and Its Indochinese Neighbors: The Enduring Logic," paper presented at a conference organized by the National University of Singapore and the Singapore Institute of International Affairs, Singapore, 5–6 November 1983; see also by the same author, "Strategic Implications of the Indochina Conflict," *Asian Affairs: An American Review*, Vol. 11, No. 3 (Fall 1984), pp. 30–32.
8. Dr. Thanat Khoman, "The Consequences for Southeast Asia of Events in Indochina," *Journal of Social Sciences*, Vol. 13, No. 1 (Bangkok, January 1976), pp. 24–25.
9. Ministry of Foreign Affairs White Book, broadcast over Voice of Free Asia, Bangkok, 14 January 1985; in Foreign Broadcast Information Service, *Daily Report: Asia and Pacific* (cited hereafter as FBIS-APA), 30 January 1985.
10. Interview with Supreme Commander Gen. Saiyud Kerdphol, *Nation Review*, Bangkok, 19 July 1982.
11. For a more detailed discussion of Vietnamese perceptions, see William S. Turley, "Hanoi's Challenge to Southeast Asian Regional Order," in Young Whan Kihl and Lawrence E. Grinter, eds., *Asian-Pacific Security: Emerging Challenges and Responses* (Boulder, Colo.: Lynne Rienner Publishers, 1986), pp. 177–200.
12. The dynamic is one of self-fulfilling prophecies in a spiral model of conflict. See Robert Jervis, *Perception and Misperception in International Politics* (Princeton, N.J.: Princeton University Press, 1976), pp. 62–67, 76–78.
13. Sukhumbhand Paribatra, "Strategic Implications," *op. cit.*, p. 30.
14. Briefing for Chat Thai Party in *Nation Review*, 13 July 1983.
15. *Lak Thai*, Bangkok, 2 May 1985; in FBIS-APA, 16 May 1985.
16. E.g., "The turmoil in Indochina has intensified political cooperation

and foreign policy coordination amongst the five [ASEAN] nations. In addition in Thailand's case it has undermined the ideological and supply bases of domestic insurgents to such a degree that Royal Thai Government political and military initiatives have succeeded in reducing the number of insurgents by more than two-thirds during the last two years. Today, insurgency is generally a dead issue in Thailand." Comments of Dr. Amnuay Viravan, Group Chairman for International Banking of Bangkok Bank, at the 1983 Hong Kong Trade Fair Conference, in *Nation Review*, 29 October 1983. See also interview with Foreign Minister Sitthi Sawetsila, *Lak Thai*, 3 January 1985; in FBIS-APA, 11 January 1985.

17. See interview with Air Chief Marshal Praphan Thupatemi, *Bangkok Post*, 3 February 1985; in FBIS-APA, 5 February 1985.
18. See interview with Squadron Leader Prasong Sunsiri, Secretary General of the National Security Council, *Nation Review*, 24 October 1985; in FBIS-APA, 24 October 1985.
19. In interviews conducted by the author during visits to Hanoi in 1983 and 1984, the deputy director of the Vietnam Committee for Economic and Cultural Cooperation with Laos and Cambodia, Pham Bao, conceded in an interview on 24 April 1984 that it was cheaper for Laos to ship goods across Thailand by train than over roads to Vietnam; but on the cultural question, all officials with whom I raised the subject contrasted the long history of Thai depredation and condescension toward the Lao and Khmer with their own more recent benign sponsorship of anticolonial revolutions and "fraternal" relations with Lao and Khmer Communist leaders. For a quasi-official Lao view, see Pheuiphanh Ngaosyvathn, "Thai-Lao Relations: A Lao View," *Asian Survey*, Vol. 25, No. 12 (December 1985), pp. 1242–1259.
20. See Turley, "Hanoi's Challenge," in Kihl and Grinter, eds., *Asian-Pacific Security*, p. 186.
21. Thach interview with Budapest Radio, 7 December 1984; in FBIS-APA, 11 December 1984.
22. *Thai Policy vis-à-vis Kampuchea* (Phnom Penh: Ministry of Foreign Affairs, September 1983), pp. 10, 55.
23. Kong Korm interview, Phnom Penh, 3 April 1984.
24. See Pisan Suriyamongkol, *Domestic Group Influence on Thai Participation in the Association of Southeast Asian Nations* (Ph.D. dissertation, University of Illinois, Champaign-Urbana, 1980), chapters VI and VII.
25. Chai-Anan Samudavanija, *The Thai Young Turks* (Singapore: Institute of Southeast Asian Studies, 1982), pp. 33–34.

26. Sarasin Viraphol, "Thailand's Perspectives on Its Rivalry with Vietnam," in William S. Turley, ed., *Confrontation or Coexistence: The Future of ASEAN-Vietnam Relations* (Bangkok Chulalongkorn University, Institute of Security and International Studies, 1985), p. 21.
27. Kramol Tongdhammachart, et al., *The Thai Elite's National Security Perspectives: Implications for Southeast Asia* (Bangkok: Chulalongkorn University, 1983), pp. 18, 19, 53.
28. Chai-Anan Samudavanija, "Implications of a Prolonged Conflict on Internal Thai Politics," in Turley, ed., *Confrontation or Coexistence*, p. 84.
29. See Sukhumbhand Paribatra, "Irreversible History? ASEAN, Vietnam and the Polarization of Southeast Asia," paper presented at the Third U.S.-ASEAN Conference on "ASEAN in the Regional and International Context, Chiangmai, Thailand, 7–11 January 1985, pp. 20–21.
30. Bangkok television service, 15 August 1986; in FBIS-APA, 19 August 1986.
31. *Bangkok Post*, 14 August 1986; in FBIS-APA, 15 August, 1986.
32. Bangkok Voice of Free Asia, 22 August 1986; in FBIS-APA, 25 August 1986.
33. See Bangkok Radio, 12 August 1986; in FBIS-APA, 13 August 1986; and Bangkok television service, 15 August 1986; in FBIS-APA, 19 August 1986.
34. Legal trade between the two countries was estimated at $12.2 million in 1985, which combined with an equal volume of illegal trade was probably double Thailand's trade with Vietnam (see below). In late October 1986, Bangkok cut the list of strategic goods banned from passage into Laos from 205 to 40 items. *Far Eastern Economic Review*, 13 November 1986, p. 60.
35. Thailand's exports to Vietnam always have exceeded its imports. The high point in trade between the two was reached in 1979, from which it plummeted to near zero in 1981–83. The reported total two-way trade volume in 1984 of barely $10 million was trivial, especially for Thailand, though it was overwhelmingly in Thailand's favor. *Direction of Trade Statistics Yearbook*, 1982 and 1985. For an assessment of the opportunity costs, see Ruangdej Srivardhana, "Implications of a Prolonged Conflict on Thailand's Economy," in Turley, ed., *Confrontation or Coexistence*, pp. 89–94.
36. See Corrine Phuangkasem, *Thailand in Southeast Asia: A Study of Foreign Policy Behavior (1964–1977)*, (Ph.D. dissertation, University of Hawaii, 1980).
37. E.g., the Ramasoon affair, which involved the secret operation of

U.S. electronic intelligence facilities on Thai soil after other American bases had been closed. See M. L. Bhansoon Ladavalya, *Thailand's Foreign Policy Under Kukrit Pramot: A Study in Decision-Making* (Ph.D. dissertation, Northern Illinois University, 1980), pp. 164–168.
38. Sukhumbhand, "Irreversible History?," *op. cit.*, p. 22.
39. Vol. 1 of a Democratic Soldiers' study document known as "Document 6601," in author's possession. Also see Chai-Anan, "Implications of a Prolonged Conflict," in Turley, ed., *Confrontation or Coexistence*, pp. 84–86.
40. Conversation with Steve Heder in Bangkok, 2 September 1983.
41. Interview with Thai Army officer by Jeffrey Race, Bangkok, 6 September 1983.
42. In 1980, the daily volume of black market transactions at the Thai-Cambodian border was estimated at $2.5 million. Banks in Aranyaprathet reported daily receipts of 30–50 million *baht*, most of which was presumed to be from black market operations. *Nation Review*, 12 September 1980. Western border relief aid—for a population of about 250,000—was $49 million in 1982. Rod Nordland, "Khmer Refugees: 'Reaching for Oars,'" *Indochina Issues*, 30 (November 1982).
43. See William S. Turley, "The Military Construction of Socialism: Post-War Roles of the People's Army of Vietnam," paper presented to the conference on "Postwar Vietnam: Ideology and Action," Institute of Development Studies, University of Sussex, Brighton, England, 9–13 September 1985.
44. See speech by Prime Minister Prem Tinsulanond, *Thailand Foreign Affairs Newsletter*, 31 December 1981, reprinted in *ISIS Bulletin*, Vol. 1, No. 1 (July 1982), p. 14.
45. For an example of such observations, see the comments of Dr. Amnuay in *Nation Review*, 29 October 1983.
46. Interview with Maj. Gen. Pichit Kullavanijaya, Bangkok, 21 February 1984. Pichit subsequently made the same point for the *Bangkok Post*, 24 April 1985; in FBIS-APA, 26 April 1985.
47. See Turley, "Hanoi's Challenge," in Kihl and Grinter, eds., *Asian-Pacific Security*, pp. 181–185 and 191–195 for discussion and evidence.
48. Interview with Thach, 25 April 1984. See also article by Jim Wolf, "Hanoi Takes Risk to Force Phnom Penh to Fight," *Bangkok Post*, 14 June 1984.
49. Gen. Le Duc Anh, "Quan doi nhan dan va nhiem vu quoc te cao ca tren dat ban Cam-Pu-Chia (The People's Army and Its Lofty Internationalist Mission in Friendly Kampuchea,)" *Tap chi Quan doi nhan dan (People's Army Journal)* (December 1984), pp. 28–43.

50. Pham Binh, "Prospects for Solutions to Problems Related to Peace and Stability in Southeast Asia," *The Indonesian Quarterly*, Vol. 12, No. 2 (Djakarta, April 1984), pp. 207, 213, 217.
51. Squadron Leader Prasong Sunsiri, then chairman of the Thai National Security Council, citing the testimony of a CPT defector who claimed Hanoi offered the CPT up to twelve battalions of Lao troops to boost the insurgency in the northeast. *Bangkok Post*, 23 December 1984; in FBIS-APA, 24 December 1984.
52. Region 2 ISOC Report, "Disclosure of Phak Mai's Plan to Take Over the 16 Northeastern Provinces," *Matichon Sut Sappada*, 13 November 1983; in JPRS, *Southeast Asia Report*, no. 11,943, n.d.
53. *The Nation*, 23 August 1985; in FIBS-APA, 23 August 1985. *Matichon*, 23 August 1985; in FBIS-APA, 29 August 1985. *The Nation*, 20 September 1985; in FBIS-APA, 20 September 1985.
54. With the insurgency rapidly disintegrating, then Supreme Commander Saiyud Kerdphol observed in a speech in 1982 that "the immediate external threat from Vietnam is not one of all-out invasion [but] rather than one of harassment and support for internal insurgents." *Thailand Foreign Affairs Newsletter* (30 April 1982), reprinted in *ISIS Bulletin*, Vol. 1, No. 1 (July 1982), p. 27.
55. Prime Minister Prem took direct control of the effort, naming himself director of the responsible agency, the Internal Security Operation Command. *Far Eastern Economic Review*, 5 March 1987, p. 26.
56. Hoang Tung, a member of the Vietnam Communist Party Secretariat, has confirmed in an interview that the troop level once reached 170,000, but claimed that number had dropped to about 100,000 in 1985. Tokyo *Kyodo* in English, 2 September 1985; in FBIS-APA, 3 September 1985.
57. Jim Wolf, "Thailand's Security and Armed Forces," *Jane's Defense Weekly* (2 November 1985), p. 980.
58. Data on Vietnam's defense spending defy reliable interpretation, but I crudely estimate Hanoi has been spending about 40 percent of domestic revenue, or roughly 10 percent of gross domestic product, on defense. Thailand's defense spending for fiscal 1986 accounted for 19.3 percent of total outlays, a decline from the average of the previous ten years, or about 4 percent of GNP. The budget for 1987 projected a slight increase for defense in absolute terms but the share was to decline further to 18.0 percent. *Far Eastern Economic Review*, 16 October 1986, p. 67; 19 February 1987, p. 26.
59. See interview with Air Chief Marshal Praphan Thupatemi, Commander-in-Chief of the Royal Thai Air Force, *Bangkok Post*, 3 February 1985; in FBIS-APA, 5 February 1985.

60. The Soviets have stationed a squadron of MiG-23s at Cam Ranh Bay since 1984.
61. Donald K. Emmerson, "The 'Stable' War: Cambodia and the Great Powers," *Indochina Issues*, 62 (December 1985).
62. Sheldon Simon, "The Great Powers' Security Role In Southeast Asia: Diplomacy and Force," in Kihl and Grinter, eds., *Asian-Pacific Security*, pp. 85–86.

8. Philippine Communism: The Continuing Threat and the Aquino Challenge*

Leif R. Rosenberger

O N FEBRUARY 25, 1986, Philippine President Ferdinand Marcos was shoved into exile after twenty years of rule, the last fourteen years under authoritarian measures. He was the victim of one of the most remarkable turns of political events in the twentieth century. Marcos's plan to round up Corazon Aquino and her key opposition followers was preempted by a mutiny organized by Armed Forces Chief Fidel Ramos and Defense Minister Juan Ponce Enrile. Ramos and Enrile said that Marcos had stolen the February 7 presidential election from Cory Aquino. Mrs. Aquino, they said, was now the legitimate leader of the country. However, it took a revolt in the military, plus mass pressure, together with timely and adroit U.S. support for the anti-Marcos forces, to dissuade Marcos from waging a bloody defense of his position. Thankfully, Marcos's journey into exile and disgrace was at least peaceful. The rise of Aquino, in turn, has opened a new chapter in Philippine history, and it creates new challenges, not only for the new government but also for the United States and the Philippine Communists.

For a country that had supported President Marcos for the past twenty years, the United States fared quite well. Since 1983, Washington was fearful that a Communist takeover in the Philippines was a distinct possibility. If the Communist Party of the Philippines (CPP) came to power, the U.S. military facilities at Clark and Subic Bay would surely be lost, perhaps giving the So-

---

*This article reflects the views of the author and not necessarily those of the U.S. government.

viet Union another strategic victory in Southeast Asia. Fear of this specter prompted Washington to put pressure on President Marcos to restore democracy by instituting sweeping political, economic, and military reforms. Instead of implementing the reforms, which to Marcos would have been political suicide, the Philippine president called a "snap election," an abortive attempt to deflect U.S. criticism and to catch the moderates weak and divided. Thanks to blatant voting fraud, President Marcos "won" the election on February 7, 1986. But Marcos lost the support of the United States and he outraged the majority of the Filipinos who wanted a fair election. Washington henceforth supported those elements in the Church and the reform elements of the military who felt Aquino deserved the presidency. While the United States did not play the major role in bringing Aquino to power, Washington's ability to persuade Marcos in the midst of the crisis to leave peacefully was not insignificant. Moreover, the United States was happy to see a moderate alternative to the Communists assume the reins of power.

The Communists, on the other hand, were anything but pleased by the sudden turn of events. The CPP had worked almost two decades to seize power once Marcos left the scene. The defection of Ramos and Enrile, and the sudden emergence of "People Power," caught the party sleeping. Aquino's victory denied the CPP what it felt was its just spoils and was a blow to the party's hopes of coming to power anytime soon. But while the sudden departure of Marcos upset the Communist timetable for seizing power, the CPP remains a formidable challenge to President Aquino's hopes for a lasting restoration of democracy in the Philippines. Aquino represents a setback for the CPP, but obviously not a permanent defeat. The party has weathered similar setbacks in the past, and is shifting its policies and efforts for the future battle. To understand the nature of the threat that the CPP now presents to the Aquino government, it is important to understand how the Philippine Communists have previously adapted to "new situations."

The main part of this chapter therefore discusses the evolution and growth of the CPP in the face of changing circumstances past and present. Patience and tactical flexibility have served the party well. In the early 1970s, the CPP was able to overcome adversity far more devastating than what confronts it today in the late 1980s. The party successfully exploited the improving revolutionary conditions growing out of the rampant corruption and gross mismanagement of the Marcos government in the early 1980s. Now with Aquino in power, the CPP is once again regrouping and assessing the changing circumstances. The latter portions of this chapter discuss the CPP reaction to Aquino coming to power, and likely tactics and strategy of the party, based on its past behavior and recent CPP statements about the future. Meanwhile, the success of President Aquino in dealing with the CPP challenge over the next year or two will be measured not only by her ability to counter the insurgency of the New People's Army, but also by her success in solving the economic, military, and political problems which the CPP and its related National Democratic Front (NDF) have successfully exploited in the past. The final section of the chapter discusses the problems which confront Aquino and possible policy options that might alleviate them, thus eroding the CPP threat. The United States can do a number of things to help the Aquino government in this time of need. But in the final analysis, the problems which confront the Aquino government must be solved by the Filipinos themselves.

I. The Rise of the CPP/NPA Threat

The CPP will have to think carefully not only about how it wishes to proceed in the face of new revolutionary situations, but also how it wishes to interact with the Soviet Union in the future. Moscow, in turn, must also review just how closely it wishes to work with the CPP now that Marcos is gone from the scene. Perhaps it would be useful to look back at the evolution of the CPP-CPSU relationship to see how both parties previously approached the ebb and flow tide of revolution.[1]

One can trace an evolution of the CPP from a Maoist, pro-Chinese Communist Party to one that, while still formally independent, shows increasing links to the Communist Party of the Soviet Union (CPSU).[2] To be sure, both the CPSU and the CPP bend over backward to obscure their links; thus, it is almost impossible to find statements in Soviet sources about the NPA insurgency, and the CPP rarely discusses Soviet activities in the

Philippines. However, by the spring of 1987 there was little question that Soviet-sponsored assistance was going to the CPP, as Soviet intelligence operations in the Philippines escalated.[3]

In the Philippines, the Kremlin faces no dilemma of choosing between pursuit of the national security interests of the Soviet Union and promotion of Communist world revolution. The USSR's perceived security interests and military objectives require an end to U.S. military superiority in the Pacific. (The placement of MiG-23 aircraft in Vietnam attests to the Soviet attempt to lessen U.S. superiority.) The removal of the U.S. military facilities in the Philippines—particularly at Clark Air Force Base and Subic Naval Base—is thus an important strategic objective.

Since the CPP/NPA has also long sought to rid the Philippines of the U.S. military presence and "U.S. economic imperialism," one might have expected the Soviets to have supported the insurgency of the New People's Army from the outset. However, the Soviets were aware that the NPA for a long time was not the equal of the Philippine military and that "capitalist stability" remained a fact of life in the island nation—i.e., that the "objective conditions" for revolution were unfavorable.

The new group of leaders around Jose Maria Sison who broke with the traditional PKP (Partido Komunista Ng Pilipinas) in 1969 to form the CPP were critical of this assessment by the Soviets. The Maoist CPP, in its zeal for armed struggle, looked to Beijing for political inspiration and material support. During this period, Beijing was chiding the Soviets for being insufficiently militant in promoting revolution. Moscow ignored these Chinese taunts and continued to support the PKP, which dutifully followed the Soviet line that capitalism was stable in the Philippines and that armed struggle was to be avoided. The PKP concentrated instead on political struggle and organizational activities.

As it turned out, Moscow was right. In the early 1970s, revolutionary conditions were absent. The CPP had a small popular following and the New People's Army was militarily weak vis-à-vis the Philippine Armed Force (AFP). The government of Ferdinand Marcos was strong, and the Philippine economy was relatively robust. Moscow continued to preach caution, patience, and political struggle. By about 1972, Sison and the CPP apparently came to the Soviet view that capitalist stability was not a short-run phenomenon in the Philippines.

It was not that the CPP suddenly "dropped" China and Maoism and became "independent." The CPP had always been Leninist in many ways, and now became even more Leninist in its outlook, especially regarding revolutionary tactics. From about 1975 to 1980, the CPP downplayed armed struggle and increasingly followed a Soviet line of political struggle.

During the latter half of the 1970s, it was China that appeared to "drop" the CPP. In their determination to counter the expansion of Soviet-backed Vietnam into Cambodia, the Chinese began to cultivate strong ties with the governments of Thailand and other members of the Association of Southeast Asian Nations (including the Philippines).[4] China also felt it necessary to improve relations with the United States in order to counter the Soviet buildup in the Far East. To reassure ASEAN and gain its support, Beijing withdrew its support from the NPA and other "national liberation movements" in the region. It also supported the presence of U.S. military bases in the Philippines as a counterweight to Soviet advances in Indochina—a position that infuriated the CPP. In 1976, Bernabe Buscayno (alias Commander Dante), leader of the NPA until his arrest in August of that year, indicated that the NPA was no longer Maoist.

Meanwhile, Moscow was busy watching for a chance to establish diplomatic relations with the Marcos government. Through the late 1960s, the internal political stability afforded by the Marcos government presented the Soviet Union with little fertile ground. As a staunch supporter of U.S. policies toward the Soviet Union, President Marcos refused diplomatic relations with the Kremlin. Although the Communist Party of the Soviet Union did enjoy the allegiance of the pro-Soviet Partido Kommunista Ng Philipinas, the PKP was relatively ineffective in the 1960s in advancing the Soviet cause in the Philippines, and was no match for the much stronger CPP. The CPSU lacked any significant links to the CPP at this time.

By the early 1970s and with the winding down of the Vietnam War, U.S.-Soviet detente and the changes in the overall East-West political climate set into motion a number of new developments.

These served ultimately to strengthen the ability of Moscow to pursue covert activities in the Philippines. Since all Soviet field agencies and representatives in the Philippines were available to support or participate in active measures, detente served to create a more favorable Soviet operational environment in the area.

In this new and more relaxed political climate, Moscow began to cultivate Philippine President Marcos in hopes of persuading him to establish diplomatic relations with the Soviet Union. Moscow did nothing covertly in the Philippines to jeopardize this critical diplomatic initiative. The Soviets were successful. In 1976, Soviet diplomatic relations with the Marcos government were established. Once diplomatic relations were set up, there was no conflict in Soviet eyes between good state-to-state relations (detente) with the Marcos government and increasing covert links with the CPP/NPA. In fact, given Moscow's past practices, it is likely that the Soviet government believed that a large diplomatic presence and concomitant increases in the Soviet cultural and trade missions were absolutely essential prerequisites for stepped-up covert links to the CPP, since most of the Soviet mission are covertly devoted to an ultimate CPP-CPSU connection—or at least the removal of U.S. bases from the Philippines.

As the 1980s approached, the CPSU and the CPP found an increasing commonality of interest in the promotion of a united front in the Philippines against the Marcos government. This was particularly evident in the labor sector, where between 1979 and 1982 new cooperation occurred between the PKP and the CPP under the auspices of the CPSU-controlled front organization, the World Federation of Trade Unions (WFTU). In October 1980, the PKP's WFTU-affiliated "Katipunan" joined forces with the CPP labor union Kilusang Mayo Uno (The May First Movement—KMU) and three ostensibly independent unions to create "Solidarity," a labor front organization opposed to Marcos's wage and no-strike policies.[5] This merger marked a significant change: for the first time, the PKP now had the full backing of the CPSU in joining overtly with the anti-Marcos opposition camp. Moreover, since the "independent" trade unions were officially affiliated with the Soviet-controlled WFTU, they were exposed to influence from Moscow. The CPP's participation in "Solidarity" was evidence that the party was no longer trying to disguise its ties to

Soviet-influenced organizations and to the PKP.

Soviet involvement in arms transfers to the NPA is more difficult to detect owing to the fact that the Philippines comprises some 7,000 islands. Nevertheless, there is evidence of Soviet involvement in at least one shipment of arms from Eastern Europe through South Yemen to the NPA.[6]

By 1981, CPP chairman Rodolfo Salas and his inner circle secretly decided to seek aid from the Soviet Union. Publicly the party denounced the CPSU as social-imperialist. This propaganda line continued as disinformation. Salas approved an operation to smuggle arms from Eastern Europe into the Philippines through South Yemen, a Soviet client state with Soviet and East German military personnel. CPP leader Salas gave the final approval to the Soviet arms shipment. Moscow used a PLO faction to deliver the weapons to the CPP/NPA. Sizable quantities of Soviet-manufactured AK-47s and Makharov pistols were put on a freighter and shipped to the Philippines.[7]

Horatio Boy Morales—former mastermind of the CPP's National Democratic Front—was a key figure in the smuggling operation. In a one-on-one courtroom interview with Boy Morales, U.S. journalist Ross Munro asked Morales why he had decided to accept aid from the Soviets. "It's a few steps removed from the Soviets," he said with a nervous laugh, adding, "but it's still considered separate, no?" The smuggling operation was done under the cover of the National Democratic Front (NDF).[8]

Long after the Soviet/PLO arms were shipped to the CPP/NPA, the CPP leadership continued to propagate the anti-Soviet party line to the rank-and-file. For instance, the CPP denounced "Soviet social-imperialism" in the August 1981 issue of its official, underground newspaper *Ang Bayan*. The NDF kept the anti-Soviet rhetoric in its party line until the year after the arms arrived.

After the details of the Soviet arms shipment became public, it became increasingly difficult to keep the anti-Soviet rhetoric in the NDF party line. In January of 1982, the CPP adopted a more pro-Soviet line, and this was included in a new draft NDF program which dropped all the Maoist jargon and the attacks on the Soviets. The CPP became openly pro-Soviet by mid-1983. *Ang Bayan* began praising developments in Cambodia, Vietnam, Mozambique, and Angola. It dropped earlier attacks on Soviet and Cuban aggression. With few exceptions, the CPP's statements on foreign policy are completely consistent with Soviet policy.[9]

The former CPP chairman Jose Sison—who had been in jail since his capture in 1977 until released by Mrs. Aquino—was angry to see this swing to the Soviets. Sison resented the fact that the Soviets refused to give him arms during the ebb tide of revolution in the 1970s, but were now giving them to CPP chief Salas during the present flow tide of revolution. Sison's vindictiveness has apparently turned him in an anti-Soviet direction.

Further evidence of the Soviet hand in Philippine developments surfaced in July 1982, when Stanislav Levchenko, former acting chief of the active measures group of the Tokyo residency of the KGB, delivered testimony on Soviet active measures before a U.S. congressional committee.[10] Levchenko said that as late as 1979, he had personally witnessed KGB officers under instructions from the CPSU's International Department "delivering money to the illegal Communist Party of the Philippines [i.e., the CPP] in bags with two bottoms. . . . [The] messenger . . . [of the] Communist Party of the Philippines . . . visited Tokyo during those years. [He] was visiting Tokyo on a more-or-less regular basis to get money from the KGB . . . I witnessed this personally. I had to help the KGB case officer . . . to be sure that Japanese counterintelligence was not surveilling the whole operation. So I was driving the car to a hotel, and the KGB case officer disappeared from the car with a heavy bag of money and went back without anything in his hands."[11]

The engagement of the Soviet bloc with the CPP was further revealed in 1983 when Carlos Gasper was discovered to be the CPP's link in a complex international funding support system. Gasper, thirty-six, an anthropologist and lay church worker, was arrested in the course of a raid by a military intelligence team on a suspected "underground" house of the CPP's Mindanao regional party committee in Davao City. Confiscated documents indicated that Gasper had traveled regularly to Europe, North Africa, the United States, Nicaragua, El Salvador, and Bangladesh to contact various groups. The documents also indicated that the CPP maintained links with Soviet-sponsored solidarity groups in foreign countries through its international liaison committee.[12]

Thus, Moscow is indirectly funding the CPP/NPA through these solidarity committees, a vehicle similar to that used by Moscow to fuel low-intensity conflict in El Salvador and Central America. These groups are an increasingly important source of funds for arms purchases by the NPA on the black market. The CPP is using its National Democratic Front to actively court the Soviets in Western Europe. Father Louis Jalandoni, a Philippine Catholic priest, is the key figure in the NDF connection to Moscow in Western Europe.[13] As the NDF's key international representative in Amsterdam, Jalandoni is successfully tying the NDF even closer to the Soviet bloc. Jalandoni was a delegate to the 1984 International Conference on Nicaragua and for Peace in Central America held in Lisbon. At the Lisbon Conference, Jalandoni conferred with Vietnam's education minister, perhaps in regard to Vietnamese shipments of old U.S. weapons to the NPA. Jalandoni's most important task has been raising funds in Western Europe for the Communist movement in the Philippines. He has been very successful. During the summer of 1981, Jalandoni provided $30,000 for travel and transportation to the CPP arms smugglers who passed through Europe on their way to South Yemen.[14]

Jalandoni has set up solidarity groups throughout Western Europe. These solidarity groups have sprung up in Sweden, Norway, West Germany, Belgium, Holland, Ireland, and several other countries. Some of the groups appear to be offspring of small, radical splinter Communist parties (e.g., Stalinist wings of Eurocommunist parties). Several of these solidarity groups sent envoys to the Philippines to see first-hand how their money was being spent by the NPA. A U.S. journalist who was visiting an NPA camp in the summer of 1985 said that a Norwegian woman was in the camp for discussions with the NPA guerrilla leader about giving financial help that would enable the NPA to obtain additional arms. Several German and Japanese radicals have also spent time with the NPA.[15]

The amount of money flowing into the Philippines from Western Europe each year is estimated by Philippine and U.S. analysts to be at least in the hundreds of thousands of dollars. Most of the money seems to be flowing from Church-related bodies in Western Europe to Communist-leaning organizations affiliated with the Roman Catholic Church in the Philippines and various Protestant churches. The liberation theology Catholics in Western Europe do not appear to be unwitting. At least some of them know precisely what is happening to their money. *Ang Bayan*, the CPP/NPA newspaper, reported that "a number of foreign church people . . . have also visited the NPA guerrilla zones."[16]

The money the CPP is receiving from abroad does not seem to have increased to the point that it is making a huge difference. But the financial support is already substantial. The CPP openly acknowledges that since 1974, it has actively been seeking and often receiving material from abroad. Philippine military authorities allege that the CPP receives about 80 percent of its agitation and propaganda funds from international funding institutions. Many of these institutions are, of course, Soviet-sponsored. According to captured documents, the CPP was using "humanitarian" organizations in the Netherlands and West Germany as a source of funds for the National Democratic Front in the Philippines.[17]

Following the August 1983 assassination of popular Philippine opposition leader Benigno Aquino, the Kremlin apparently reassessed revolutionary conditions in the Philippines. Judging from an analysis of the Soviet press since the Aquino assassination, one would say the Soviets feel that revolutionary prospects are improving. Seen through a Soviet ideological lens, the social and economic pressures for change are building; opposition to the "Marcos-U.S. dictatorship" is increasing; and the ability of the Marcos regime to cope with the rising tide of Communist insurgency and opposition sentiment among non-Communist groups is declining.

The Soviets moved quickly to exploit the anti-Marcos, anti-American sentiment that gathered momentum after the Aquino assassination. The Kremlin increased the size of the Soviet mission in Manila (from about sixty to ninety) in late 1983 and early 1984. The character of the mission changed: energetic Soviet covert operatives replaced tired old diplomats. For example, the experienced KGB official Boris Smirnov—who in 1976 had successfully passed off in Tokyo what was reputed to be the "last will of Zhou Enlai"—was assigned to Manila on April 12, 1984, as first secretary of the Soviet Embassy. A few weeks after his arrival in Manila, a bogus questionnaire "from the U.S. Information Service" was distributed among leading Filipinos, seeking sensitive information on subjects such as their political leanings and military experience. This anti-American disinformation operation was reportedly successful in generating Filipino outrage at the "impertinence" of the allegedly American questionnaire.[18]

Moscow invariably hides many of its sensitive activities with the CPP under the cover of cultural, sports, or economic exchanges and visits. These people can travel to places that are off limits to Soviet "diplomats." According to an official of the Philippine government, in the last ten months of 1984 there was an upsurge in the number of scientists and cultural troupes from the USSR and Eastern Europe interested in attending international conferences or performing in the Philippines. These contacts have taken place at both the official and the people-to-people levels.[19] In addition, the Soviets advance their interests through the "parliament in the streets." The PKP, almost totally controlled by the CPSU, is a partner of the CPP in many demonstrations and rallies, and members of both Philippine Communist parties work together (along with other political groupings) in the National Democratic Front.

From November 30 to December 4, 1984, the Soviet-controlled World Peace Council (WPC) and its Philippine affiliate, the Philippine Peace and Solidarity Council, co-sponsored the first International Conference on Peace and Security in East Asia and the Pacific. The conference was held at the University of the Philippines, a central location for National Democratic Front activity. It therefore afforded the Soviets a golden opportunity to use their WPC affiliate, the Soviet Committee for the Defense of Peace, and its PKP proxy to strengthen ties with the CPP, and to fan opposition to the U.S. bases and "American Imperialism."[20]

This WPC-sponsored conference coincided with a wave of protests and demonstrations throughout the Philippines. *Pravda* quotes an alleged participant in one of the meetings: "There were more than 5,000 people. We gathered in front of the American base at Clark Field, the largest in this region of the world. From here the bombers flew to Vietnam. The Americans call the entrance to the base the 'Gate of Friendship,' but we call them the 'Gates of Hell,' since nuclear death lurks behind them. We demanded the removal from our land of all 23 US bases, including Clark Field and Subic Bay."[21]

As is usual in most cases of Soviet involvement in low-intensity conflict, it is the Soviet Union that is being wooed by the CPP/NPA for aid. Moscow in no way is trying to push itself at the CPP. The new generation of CPP rulers are pragmatic and opportunistic. Maoism is a vague memory. The CPP is not hesitant to accept substantial amounts of aid from the Soviets. During 1985, the success of the CPP/NPA has made it exceedingly difficult to keep the insurgency going strong without increasing amounts of external support. The success of the NPA is attracting increasing numbers of new recruits. The NPA is also moving in larger units. All of this is creating a soaring demand for arms, food, and equipment. There is evidence to suggest the NPA now has enough potential recruits to increase in size if it had the money to equip and support them. If the economy was strong, then the CPP/NPA could continue to rely heavily on internal sources of support. But a huge debt and economic depression in the Philippines—the very factors which help to produce so many potential recruits—make NPA tax collecting increasingly less lucrative.

Contrary to some reports, it is unlikely that there is debate going on in Moscow over whether to support the CPP/NPA or to "drop" them. The CPP/CPSU connection is well established. Nevertheless, a debate does exist in the Kremlin over how to support the CPP/NPA insurgency at the moment. The CPP/NPA is not hiding the fact that it would like the Soviet Union to provide stepped-up, direct military support to the insurgency.

Some Soviets probably prefer to continue the ongoing policy of relatively low-level military support and high-level financial support. These cautious Soviet leaders believe this approach was successful in Nicaragua and will be successful in the Philippines. They argue that if the Soviet Union became blatantly involved in military support, there would be an unacceptable "blowback." That is, the United States would then view the Philippine insurgency as an East-West issue and intervene militarily.

The more bullish Soviets probably disagree. They argue that the CPP/NPA is in desperate need of more arms and ammunition. If Moscow fails to supply these, the CPP/NPA may go elsewhere for military support, thus causing Moscow to lose a potentially key pro-Soviet state in Southeast Asia. These bullish Soviet leaders also argue that when the shipment of AK-47s in 1981 and 1982 was revealed, the United States failed to react. The debate will probably be won by the more cautious Soviet leaders, who argue that the Soviet Union can best advance its interests by providing the optimum amount of support to the CPP/NPA, while at the same time hiding its hand from the United States.

II. CPP Reaction and Response to Aquino

The incredible and sudden fall of Ferdinand Marcos and the rise of Corazon Aquino to power caught the Communist Party of the Philippines totally by surprise. And while most Filipinos were euphoric, the departure of Marcos was a blow to the CPP. The CPP, which takes pride in its ability to mobilize protests and demonstrations, was left behind by the "people's power" and the military rebels backing Cory Aquino. Moreover, Aquino's electoral claim to the presidency was in no way due to the efforts of the CPP. The CPP's boycott of the February 7 election meant that it could take no credit for the "restoration of democracy" in the Philippines. And the CPP frankly had lost a "friend" in Marcos: Marcos had become the focal point of its propaganda and recruiting effort. Prior to the February 22 revolt, even some moderates felt that they would ultimately have to turn to violence and unite with the CPP if they were ever going to get rid of the intransigent Marcos. With Marcos now gone, such alliances became no longer logical or necessary.

At a February 25, 1986, Politburo meeting, the CPP mood was reportedly gloomy. Communist leaders seemed "numbed" by the situation and "dwarfed" by the turn of events. Until Marcos fell, CPP members were bullish on their chances for success. With Marcos gone, the party became pessimistic. While such pessimism is probably overstated, the new situation did mean that the CPP had to rethink its tactics and strategy in the wake of a new reality. At the minimum, it will take time for the party to develop a consensus among its leaders on both tactics and strategy.

The Soviet Union was also caught by surprise by the fall of Marcos. The USSR was the only government to congratulate

Marcos on his election, and it supported him to the bitter end. In the past, this policy of good relations with Marcos had paid dividends by allowing Moscow to increase the size of its embassy, thus enabling the Soviets also to expand their covert ties to the CPP. But like the CPP, Moscow was out in the cold, with no direct links to the liberal-minded Aquino. Traditionally, the Soviets are much more comfortable dealing openly with right-wing capitalists (and covertly with Communists) than with liberals.

Meanwhile, the Soviet Union, which has been advising the CPP to be sensitive to the ebb and flow tide of revolution, will no doubt caution the party to be careful and to avoid an exclusive reliance on armed struggle, at least for a while. While Marcos was in power, the CPP could successfully argue that its military arm, the New People's Army, was pursuing armed struggle in order to get rid of Marcos and "restore democracy." Now that Marcos is gone and "democracy has been restored," this propaganda line will not sell. Propaganda directed against the Armed Forces of the Philippines (AFP) also will not sell, since Ramos and Enrile—the leaders of the AFP—had allied themselves with a democratic cause.

The fall of Marcos and rise of Aquino created a number of problems which the CPP was forced to address almost immediately. Aquino's enormous popularity was the first problem facing the party. Her popularity threatened to weaken CPP united front structures (such as the NDF), especially in the urban areas.[22] Aquino's popularity also threatened to spark a split in the CPP. Many NDF members were not hard-core Communists and had joined the NDF in order to get rid of Marcos. Now that Marcos was gone, these NDF members were vulnerable to appeals from the Aquino camp. Aquino's popularity also threatened to spark a split in BAYAN, another key CPP front organization. Some factionalism had already occurred in BAYAN during the February 7, 1986 election. A number of local BAYAN chapters broke with the CPP leadership's call for a boycott and actually campaigned for Aquino.[23] When Aquino assumed power, some members of BAYAN and another front group, the National Alliance for Justice, Peace, and Freedom, advocated political action rather than armed struggle. Still another front group, the Nationalist Youth, proposed a dialogue with the Aquino government.[24] And even in the CPP itself, divisions seemed to be developing. In its March 1986 analysis of the new situation, differences were apparent over the necessity of armed struggle.[25] Aquino's cease-fire and amnesty offer threatened to fan this factionalism by eroding rank-and-file insurgents, and even some local CPP and NPA leaders.[26] Aquino's appeal threatened to attract the soft-core or socially discontented party members, many of whom are not true Marxist-Leninists.[27] And finally, some CPP and NPA members might leave the Communist camp because of Aquino's stated commitment to reform and revitalize the AFP, and make it an effective fighting force against the CPP/NPA.

The CPP Recovery

The worst fears of the CPP leadership (and Aquino's hope), that the fall of Marcos would encourage most of the NPA fighters to come down from the hills and surrender, did not occur. The urge to revolt at the grass roots was as strong as ever. In fact, insurgency-related violence, almost all initiated by the NPA, actually increased slightly during Aquino's first one hundred days, with an average of eleven persons killed daily during the period compared to ten per day from January 1 to February 22, according to AFP chief Ramos.[28] By mid-1986, about 800 people had died in AFP/NPA fighting since Aquino took office in February 1986.[29] In addition, the number of NPA rebels surrendering nationwide during Aquino's first 100 days came to a disappointing 1,652, of which only 102 were listed as NPA regulars. The rest included 489 persons described as rebel activists and 1,061 members of the Communists' "Mass Base," meaning civilian supporters in rebel-controlled areas.[30] In southeastern Mindinao, only about 200 NPA rebels had surrendered by June 1986 out of a total of about 8,600 operatives.[31] On June 3, 1986, on the eve of Aquino's first one hundred days as president, U.S. Deputy Assistant Secretary of Defense Richard Armitage said that the NPA was growing stronger and more violent. According to Armitage, "The military situation is serious, and getting worse with the Communists enjoying the initiative and assuming de facto control in areas where government influence has ruled over the years."[32] Armitage concluded by saying that recent NPA actions "leave little doubt in our mind that, at the end of the day, military action will be required to defeat the insurgents."[33]

The Peace Offensive

Immediately after Aquino came to power, the CPP labeled her government "reactionary" in character. In a paper entitled "The New Situation and the Immediate Task," the CPP executive committee ordered: "If the government and the military calls for negotiation, surrender or cease-fire, do not entertain these. . . . We shall oppose these calls." Then on March 12, the executive committee softened its line and stated that, even as the situation in the urban centers warranted a closer consideration of the cease-fire plan, "There is not enough basis for us to actually enter into such an agreement."[34] During the rest of March and most of April, CPP members continued to debate the decision to boycott the February 7 election. In addition, confusion over how to deal with the Aquino government reportedly delayed an official CPP response to Aquino's appeal for talks.[35]

But in late April or early May 1986, the fifteen-man CPP Politburo reportedly met in Luzon for the first time in years.[36] Party leaders acknowledged that the decision to boycott the February 7 election was the CPP's biggest political blunder in the history of struggle, because it isolated the party and its leftist allies from the popular revolt that eventually brought down Marcos and elevated Aquino to power.[37]

Former party chief Jose Maria Sison said on June 4, 1986, that those CPP members who carried out the rigid boycott have already done some "self-criticism." He maintained that the error or mistake had been tactical, however, rather than strategic, and thought it would not lead to a purge in the party. Nevertheless, a number of unconfirmed reports say that CPP chairman and military commander Rodolfo Salas and secretary general Rafael Baylosis face ouster for having advocated the boycott. (Salas was captured in Manila by the AFP on September 29, 1986.) Most importantly, the CPP Politburo decided that while the "armed struggle should continue," tactics other than military assaults can place the Communists in a "position of influence." In the nonmilitary realm, the Politburo unanimously decided to explore the possibility of cooperating with Aquino's call for cease-fire talks.

Whether or not the CPP wants to become a legal party and vie for political power in elections is unclear. But the Politburo decision, in turn, paved the way for the June 5 announcement that Stur Ocampo, a fugitive former journalist, would be the official CPP representative for "preliminary talks" on a cease-fire between the NPA and the AFP.[38] Asian specialist Larry A. Niksch believed that a CPP decision to negotiate probably would serve several tactical ends. These included:

- Display political flexibility to recent critics in CCP front groups.
- Preclude local NPA commanders from negotiating separately.
- Use talks to play up alleged divisions between Aquino and her military advisers.
- Use talks to publicize "nationalist" issues.[39]

The CPP has been quite candid about the ground rules for a permanent cease-fire. Its main condition is that the AFP troops must remain in their barracks, including the Civilian Home Defense Forces (CHDF).[40] According to Jose Sison, the party was not really interested in a temporary cease-fire, which he said the AFP wanted as a "breather" to retrain, reorganize, and consolidate its forces. Instead, the CPP still wants a "lasting cease-fire" in connection with the formation of a coalition government. Sison says such a coalition would be one of the CPP's major conditions for agreeing to a lasting truce. Additional conditions reportedly included the right to maintain the integrity of the NPA as part of the CPP and recognition of the new "national revolutionary army" under the coalition government.[41]

The CPP also has indicated an interest in a permanent cease-fire if it resulted in "liberated areas." If AFP soldiers agreed to stay in their barracks, this position would allow the party to consolidate its control over areas where it is dominant or has an overwhelming presence. At a minimum, it would help the CPP prevent or limit the erosion of NPA members and the rural mass base, at least in the areas it controls.[42] At the same time, the CPP would undoubtedly continue its front-building activities in AFP-controlled areas, and according to one observer, "the CPP would likely employ intimidation of local officials and possibly assassinations and other low level violence as part of organization building—on the assumption that the government would not react to such low level activity."[43] Some NDF spokesmen have also suggested that the resolution of the land question (implementation of a genuine land reform program) and the removal of the U.S. military facilities are also terms for a permanent cease-fire, but former NDF chief Boy Morales has said that these issues are, more likely, terms for a political settlement with the Aquino government and not terms for a cease-fire.[44]

CPP Propaganda

At the same time that the CPP was involved in cease-fire talks, it continued its propaganda offensive. Aquino is still too popular to attack personally. Instead, the CPP sought and will continue to seek to drive wedges between the hawks and doves in Aquino's cabinet. It has praised the "progressive" Aquino and her leftist-leaning cabinet members such as Labor Minister Sanchez, Executive Secretary Joker Arroyo, and Special Counsel Saguisag. It blasted the "warmongering Enrile-Ramos clique" and U.S. "Imperialism." The NDF has charged that the Reagan administration is prodding Aquino to "modernize" the AFP and launch unrelenting counterinsurgency operations.[45] The NPA is portrayed as a defensive organization, protecting the masses and the revolutionary movement from the provocative, offensive attacks and acts of terrorism of the AFP.[46] The NPA urges President Aquino to "rein in and discipline her warmongering generals," who are allegedly trying to sabotage Aquino's just efforts at finding a peaceful solution to the seventeen-year-old guerrilla warfare struggle.

III. Aquino in Power: Challenges and Policy Options

When President Aquino assumed power, on February 25, 1985, she inherited staggering problems and pressures. She faced a severe economic crisis, a mandate for political, economic, social, and military reform. The CPP/NPA, in turn, was exploiting these "revolutionary conditions" and enjoying increasing success in its insurgency against the ill-equipped Philippine armed forces. The military, in turn, was divided, weakened, and demoralized by frequent charges that General Ver, the former chief of staff, and other key military leaders were behind the 1983 slaying of Cory Aquino's husband, popular moderate leader Benigno Aquino. While the basic problems facing Aquino are obvious, there are no easy solutions. In her first ten months, Aquino made a number of sound moves. She did away with the major state-run monopolies (crony capitalism) which had been sapping the Philippine economy. She improved morale of some middle-level AFP officers by getting rid of General Ver's clique of overstaying generals. Her highly successful visit to the United States and Vice President Laurel's follow-up visit also boosted the U.S. business community's confidence in reinvestment.

However, Mrs. Aquino's presidency has not lived up to everyone's expectations, particularly in the security arena, where her hopes that the Communists would lay down their arms and come down from the hills once Marcos left have not been realized. While the nonmilitary aspects of a counterinsurgency are ultimately decisive factors, a successful counterinsurgency must also include at least a credible military element, if only to deter the enemy from believing it can obtain political power primarily through armed struggle. Mao once said that power comes out of the barrel of a gun. The Chinese leader also said that revolution is not a tea party. This militant image of the NPA is alien to Aquino. Consequently, Mrs. Aquino's policy toward the NPA had some cracks, especially in the security arena. For instance, one of Aquino's first questionable moves was to give amnesty to former CPP Chief Jose Mara Sison, former NDF Chief Bay Morales, former NPA Chief Commander Dante and over 400 other political prisoners, some of whom were members of the CPP. The CPP, which has been in desperate need of more brain power, thus received an unexpected shot in the arm. Both Defense Minister Enrile and Chief of Staff Ramos publicly criticized Mrs. Aquino's actions.

Aquino's conciliatory and magnanimous policy toward the NPA did not last indefinitely, however. The Philippine president indicated in March 1987 that her peace initiatives had failed and the time had come for "military victory" over both Communist insurgents and right wing terrorists.[47] More Aquino acts of this nature are needed to win the complete support of the AFP and unify the non-Communist Filipinos against the rebels. In addition, Mrs. Aquino realizes that a successful counterinsurgency must contain primarily nonmilitary components. (Marcos never understood this reality, and as a result the AFP under Marcos was consumed with trying to kill rebels while almost totally neglecting and occasionally terrorizing the people it was supposed to protect.) To her credit, Aquino also realizes that the AFP must develop popular support. In addition, President Aquino wants to come up with economic and social counterinsurgency programs to "win the hearts and minds" of those poor Filipinos who are tempted to lean toward the CPP/NPA as alternatives to the existing socioeconomic system. Unfortunately, Aquino has yet to formulate a comprehensive civic action program that would be consistent with this vision—let alone implement one. Still, there is some reason for optimism. The AFP has recently submitted a framework for a comprehensive counterinsurgency program, one that stresses social, economic, and political components. But first, President Aquino and elements of the AFP must resolve their disagreements and mutual suspicions. This is not to suggest that Aquino has not had her share of positive actions that deserve the full and unequivocal support of the U.S. government. To her credit, she has taken numerous actions which help to restore democracy in the Philippines. For instance, she appointed a constitutional commission that included five KBL members. A referendum on this constitution won by an overwhelming majority in February 1987. She has appointed an independent supreme court and she has kept the Bill of Rights. She has appointed new members to the Commission on Elections. She has scheduled legislative elections in May of 1987 and later in 1987 local elections are due to take place. And late in the fall of 1986, Mrs. Aquino dropped most of her left-leaning officials from her cabinet, including Labor Minister Sanchez, who had antagonized AFP leaders. These moves are constructive.

But the record also shows actions by Mrs. Aquino that were controversial and drew public criticism from many observers. For instance, Aquino initially appointed a number of left-leaning officials to key cabinet positions, a decision that prompted numerous disagreements from members of the AFP leadership. And during Aquino's more liberalized political atmosphere, the CPP took the opportunity to expand their front-building activities in the cities. Some of her other activities were errors of commission rather than omission vis-à-vis the CPP/NPA. For example, on May 1, 1986, Mrs. Aquino made her first Labor Day speech with former CPP Chief Jose Maria Sison and former NPA Chief Commander Dante at her side.[48] In addition, Aquino has angered AFP leaders by opposing military initiatives and boldness, while not yet moving as decisively against reports of NPA ambushes and terrorism.[49] And while she talked of amnesty and reconciliation for the NPA, she demoralized elements of the military because of the ongoing investigation by the Commission on Human Rights regarding past and present abuses by the AFP. Meanwhile, the AFP continues to be handicapped by inadequate resources, partly because of U.S. congresionally-mandated reductions in U.S. military aid as well as Aquino's recent 14 percent cut in the 1986 defense budget. Under such circumstances, AFP corruption and military abuses (which fanned the insurgency) may not be immediately eradicated. Moreover, President Aquino's expectations for the ceasefire talks that ended in early February 1987 may have demonstrated a misunderstanding of CPP strategy and tactics. The CPP offensive was undoubtedly designed to take advantage of the leftists in Aquino's cabinet who embraced the December 1984 "Declaration of Unity," a document signed by twelve of the major moderate leaders opposed to Marcos.[50]

This key document reflects what the leftists in Aquino's cabinet want. It called for the successor to Marcos to seek to remove "foreign" (i.e., U.S.) military bases, legalize the CPP, give unconditional amnesty to political prisoners, and review economic agreements with foreign countries (i.e., dismantle "U.S. economic imperialism"). Former Labor Minister Blas Ople argued that Aquino's soft line toward the CPP will enable the CPP to impose a coalition government on her within two years.[51]

Many observers also question her decision to take actions against KBL supporters, whose only crime was to vote for her opponent. She summarily fired KBL governors, mayors, and local officials (many of whom were fairly elected by the people) and replaced them with many inexperienced "Officers-in-Charge"—or

OICs. Such practices are examples of "an incomplete democracy." Similarly, during Aquino's interim period she abolished the Philippine legislature and unilaterally proclaimed a provisional constitution that gave her powers at least as great as those of the deposed President Marcos.[52] Blas Ople argued that the abolition of the legislature, whose members were also elected by their constituencies, and the dismissal of thousands from the civil service on partisan grounds have tended to polarize the country and exacerbate unrest. And while the KBL political party was certainly fraudulent in its electoral tactics, it still represents a strong minority of Filipinos.

Overall, President Aquino has done a number of constructive things. But her willingness to be magnanimous and conciliatory to the CPP/NPA while downplaying their "human rights abuses" is inconsistent with many of her actions against elements of the KBL and the AFP. A far better policy would be one that is evenhanded and seeks to unify all the democratic elements of Philippine society.

IV. Proposals for Stability

Economic Stabilization

Attacking the economic problems in the Philippines will go a long way to eroding the CPP/NPA threat and reestablishing peace and stability in the country. In this regard, a number of things can be done to get the economy moving in the right direction. First, all remnants of Marcos's "crony capitalism" must continue to be dismantled. Government control over the private sector needs to be ended. As Aquino's finance minister, Jaime Ongpin, put it: "I think the government should get out of business completely. Privatize everything. . . ."[53] Barriers to foreign investment also need to be eliminated if a free enterprise system is to be unleashed in the Philippines. Then the country would be better able to take advantage of programs like the so-called Baker Plan, which calls for massive multilateral efforts to restructure developing countries' debt burdens.

But considering the economic mess in the Philippines, the

CPP/NPA threat, and the critical importance of the Philippines to allied security plans, would not a Marshall Plan be more appropriate? In other words, would outright grants in aid, rather than simply rescheduled loans, be more effective? Or can the country absorb such aid effectively? For example, the Philippines needs instant interest relief from EXIMBANK on the huge loan for the current nuclear power plant which cost over $2 billion. Manila also needs assistance on payment for FMS credit maturities at this time. In addition, humanitarian food aid is needed for Negros and other severely depressed areas of the country. And finally, Washington should do everything possible to expedite the repatriation of the billions of dollars Marcos stole from his fellow Filipinos.[54]

At a minimum, the United States needs to work closely with the World Bank, the International Monetary Fund, and the Asian Development Bank to lessen the debt burden in Manila. Even better would be a U.S. effort to work with countries such as Japan to develop multilateral grants for immediate budget relief in the Philippines.

Special attention should be given to helping small- and medium-scale farmers who need technical assistance to improve production without too many costly imported products. Crop diversification needs to be started at once. Manila also needs grants to start up a private investment fund for agri-based small- and medium-scale enterprises in rural areas.

Of course, a revitalized Philippine economy cannot take place without an upsurge in exports. But Philippine exports need markets. What better way to demonstrate U.S. support for a Philippine economic recovery than to open up U.S. markets to Philippine exports? The Philippines' share of the U.S. garment imports is only 2.5 percent. Compare this figure to recent evidence on other Asian countries' exports to the United States: Communist China's 9.9 percent, Taiwan's 13.7 percent, Hong Kong's 10.5 percent, Korea's 11.2 percent, and Japan's 7.5 percent.[55] Surely Manila deserves a better deal. The same holds true for the Philippines' share of U.S. sugar imports. Presently, the Philippines' sugar quota is 13.5 percent. Compare this figure to the 17.6 percent which the Dominican Republic enjoys and the

14.5 percent which Brazil has. The U.S. Philippine sugar quota used to be 25 percent of the U.S. market.[56] At a minimum, the Philippines deserves a most favored nation treatment for sugar equal to the benefits included for the Dominican Republic in the Caribbean Basin Initiative legislation that was introduced in 1982. Increasing the Philippine sugar quota in turn would serve to increase employment in the depressed sugar-producing areas such as Negros, where malnutrition and other maladies are attractive targets for CPP/NPA exploitation.[57] In the words of the Philippine Ambassador to the United States, Emmanuel Pelaez, "If the United States would approximate the assistance it gives to South Korea by opening its markets to us, it would be wonderful." The ambassador continued, "We would like the United States to help us in the way we need to be helped, rather than how you feel we should be helped."[58]

Revitalizing the AFP

The Philippine armed forces are another clear priority. Revitalizing the military is under way. General Ramos has rid the AFP of many of the cronies whom Marcos and General Ver had promoted to generals. And the AFP boosted its stock by helping to force Marcos out of power. But to the average farmer in the remote areas of the country, not much has changed. The AFP still suffers from an image problem and is criticized for "abuses." To counter CPP/NPA propaganda, the AFP must win back the support of farmers and workers. One initiative would be a more effective and comprehensive civic action program, which brings food and supplies to remote areas of the country ignored under Marcos.

In addition, the Aquino government must reverse its constraints on the military budget (the lowest in ASEAN) and instead allocate more resources to the ordinary soldiers. The military budget also must be increased to improve critical communications and troop support.[59]

Finally, as her talks with the rebels stalemated in the summer and fall of 1986, rising pressure on Mrs. Aquino from her military commanders, and Secretary of Defense Enrile, seemed to prompt her into a tougher policy. In remarks at Harvard University during her U.S. visit, she declared that if nonviolent means fail, as "a final option we will have to use force."[60] And when NPA military commander Rodolfo Salas was captured on the streets of Manila on September 29, Mrs. Aquino followed Enrile and Ramos's suggestions that he be prosecuted.[61] In late October, Mrs. Aquino and Mr. Enrile barely avoided a major cabinet crisis over her policies toward the Communists and other issues.

Restoration of Democracy

The Aquino government came to power as a popular alternative to the corruption of the Marcos regime. Many observers hoped that the government would restore democracy in the Philippines as soon as possible. The government remains a "revolutionary" government. President Aquino ignored elements of the interim constitution, and fired elected people just because they were members of the KBL. Instead of replacing the KBL by fiat, local elections in the rural areas should be held as early as possible. Much will depend on how the new constitution is implemented. Then, should new elections be held under constitutional auspices and Mrs. Aquino or Mr. Laurel win, they could institutionalize their political legitimacy, allow for a reconciliation with fairly elected KBL members, and facilitate a restoration of real democracy in the Philippines.

V. Future Prospects

Crony capitalism, politicized, unprofessional armed forces, and a corrupt authoritarian political system all contributed to rising revolutionary conditions in the Philippines which, in turn, enabled the CPP/NPA to grow. The Aquino administration has made progress in removing corruption from politics, although the system remains largely authoritarian. Further progress in these areas can be expedited, given the political will. But the economic stagnation in the country is not something that can be solved quickly, even if the most astute stabilization and investment policies are followed. For sound economic growth, investment, trade, and aid policies will take time to jell. Further, their impact will take years to trickle down to the peasants and workers. In the interim, the CPP/NPA will almost certainly be able to exploit the "persisting economic problems" and appeal to the "downtrodden." Even if the economy begins to show signs of improvement, the CPP/NPA is likely to strike at economic targets as a way to sabotage the improvements.

The CPP/NPA threat, therefore, will not go away any time soon. The low-intensity conflict can be controlled, however, if the Aquino government pulls out of its ambivalence and takes a hard line toward those CPP/NPA members who refuse to accept Manila's amnesty and rehabilitation offers. If, concurrently, the AFP is allowed to seek out and eliminate CPP/NPA military strongholds, the military threat would ultimately become manageable. Given a decreased CPP/NPA military threat, the Aquino government could afford to put more of the budget into the economy so as to rekindle long-term economic growth and job creation. But if the soft line toward the CPP/NPA military insurgency continues, then the prospects for economic revitalization and political stability are remote, and the opportunites for CPP/NPA gains will grow. Moreover, the government may face a revolt within the military.

An analogy might be made to a house infested with termites that catches fire. Termites arguably may have put the house into such a state that when a small fire started, it quickly spread throughout the building. Once the fire is put out, one can argue about whether to fix up the old house or build a new one. But first, the fire itself must be put out, or it may spread to other houses. In the Philippines, the government must allow the armed forces to put out the fire that the Communists are waging in the country. To minimize these fires is to guarantee an outcome where nothing can be rebuilt, and where a Khmer Rouge–type regime will be perfectly content to rule over the ashes.

Notes

1. For a more comprehensive discussion of the evolution of Philippine communism, see Leif R. Rosenberger, "Philippine Communism and the Soviet Union," *Survey*, Vol. 29 (Spring 1985).
2. For an alternative interpretation that the CPP/NPA is totally independent, see David A. Rosenberg, "Communism in the Philippines," in *Problems of Communism* (September–October 1984).

3. See, for example, the evidence on growing Soviet embassy, intelligence, and weapons-funneling operations in Tom Breen, "U.S. Monitoring Soviet Role in Philippine Insurgency," *Washington Times*, 24 March 1987, pp. 1A, 6A.
4. For more on the dynamics of Soviet-Chinese competition in Indochina, see Leif R. Rosenberger, "The Soviet-Vietnamese Alliance and Kampuchea," *Survey*, Vol. 27 (Autumn-Winter 1983).
5. *International Press*, 28 November 1983, p. 703.
6. Ross H. Munro, "Dateline Manila: Moscow's Next Win?", in *Foreign Policy*, 56 (Fall 1984), p. 186. Also see Tom Breen, "U.S. Monitoring Soviet Role in Philippine Insurgency," *Washington Times*, 24 March 1987, pp. 1A, 6A.
7. Ross Munro, "The New Khmer Rouge," in *Commentary* (December 1985), p. 36.
8. *Ibid.*
9. *Ibid.*, p. 37.
10. "Soviet Active Measures"—Hearings before the Permanent House Select Committee on Intelligence (Washington, D.C.: gro, 1982), 13–14 July 1982, pp. 166–177.
11. *Ibid.*
12. See Foreign Broadcast Information Service, *Daily Report: Asia and Pacific*, 28 April 1983, pp. P1–P4, and *Manila Bulletin Today*, 3 April 1983.
13. For a comprehensive discussion of Mr. Jaladoni's activities, see Sol Juvinda, "SI Louie Jalandoni, NDF International Representative, Sa Paris," *Malaya Sunday*, 2 February 1986.
14. Ross Munro, "The New Khmer Rouge," *Commentary* (December 1985), p. 37.
15. *Ibid.*
16. *Ibid.*
17. *The Manila Times Journal*, 29 March 1983.
18. *Business Times*, Kuala Lampur, Malaysia, 5 December 1984.
19. *Ibid.*
20. *Pravda*, 4 December 1984, p. 1.
21. *Ibid.*
22. Larry A. Niksch, *The Communist Party of the Philippines and the Aquino Government; Responding to the New Situation*. Prepared for a conference on "Crisis in the Philippines," sponsored by the Washington Institute for Values in Public Policy, 30 April–1 May 1986, p. 14.
23. *Ibid.*
24. *The Philippine Daily Enquirer*, 4 March 1986.
25. *Veritas*, 3 April 1986.

26. Niksch, *The Communist Party of the Philippines*, p. 12.
27. *Ibid.*
28. *Washington Post*, 5 June 1986, p. A 30.
29. *Washington Post*, 25 May 1986, p. A 27.
30. *Washington Post*, 5 June 1986, p. A 31.
31. *Washington Post*, 5 June 1986, p. A 21.
32. *Washington Post*, 4 June 1986, p. A 1.
33. *Washington Post*, 4 June 1986, p. A 1.
34. *Philippine Daily Enquirer*, 14 May 1986, pp. 1–2.
35. *Washington Post*, 6 June 1986, p. A 29.
36. *Philippine Daily Enquirer*, 14 May 1986, pp. 1, 2.
37. *Washington Post*, 6 June 1986, p. A 29.
38. *Washington Post*, 6 June 1986, p. A 29.
39. Niksch, *The Communist Party of the Philippines*, p. 19.
40. *The Manila Evening Post*, 3 May 1986, pp. 1, 3.
41. *Washington Post*, 6 June 1986, p. A 29.
42. Niksch, *The Communist Party of the Philippines*, p. 20.
43. *Ibid.*
44. *Business Day*, 5 May 1986.
45. *We Forum*, 6–12 May 1986, pp. 3, 15.
46. *The New Philippine Daily Express*, 14 May 1983, p. 3.
47. Seth Mydans, "Aquino Demands Military Victory over Insurgents," *New York Times*, 23 March 1987, pp. A1, A8. And by March 1987, senior Pentagon officials were arguing that rebel strength had increased by nearly 10 percent, to 24,430, since Mrs. Aquino took office.
48. *Business Day*, 7 May 1986, p. 4.
49. *Business Day*, 16 May 1986, p. 19.
50. *Washington Post*, 27 December 1984, p. 10.
51. *Business Day*, 16 May 1986, p. 14.
52. *Los Angeles Times*, 26 March 1986, p. 1.
53. Paul J. Gigot, "Manila's Economic Revolutionary," *Wall Street Journal*, 5 March 1986, p. 32.
54. On 6 October, U.S. Under Secretary of State Michael Armacost suggested that if Marcos returned the money he stole, the Aquino government ought to consider allowing him to return home. Remarks at Tufts University, Medford, Mass., 6 October 1986.
55. Ambassador Emmanuel Pelaez, "The Bottom Line: Freedom, Justice and Dignity," speech delivered at the Washington Institute for Values in Public Policy, 1 May 1986.
56. David Reinah, "Problems in the Philippine Sugar Industry," paper submitted on 30 April 1986 as part of the Conference on the Philip-

pines and U.S. Policy at the Washington Institute for Values in Public Policy, p. 9.
57. See, for example, Clayton Jones, "In Philippines, Sugar Barons Seek Ways to Combat Insurgency," *Christian Science Monitor*, 24 September 1986, pp. 10–11.
58. Pelaez, "The Bottom Line," *loc. cit.*
59. Richard D. Fischer, Jr., "The US Can Help Manila Rebuild Philippine Democracy," *Backgrounder* (The Heritage Foundation), 24 March 1986, p. 4.
60. *Washington Post*, 21 September 1986, p. A 24.
61. Clayton Jones, "Aquino Gets Tough on Communists," *Christian Science Monitor*, 2 October 1986, p. 9.

9. Conclusion: Opportunities for Deescalating East Asia's Conflicts

Young Whan Kihl and Lawrence E. Grinter

THE preceding analyses of emerging trends, conflicts, and policy opportunities in East Asia and the Western Pacific, while emphasizing local and regional conflicts, also show a pattern of action-reaction by the United States and the Soviet Union, contributing to their force escalations in the region. Soviet power projection in Asia in recent years has undoubtedly been motivated by the desire to establish its legitimate claim as an Asian-Pacific power. But it also has occurred in response to the American retrenchment in the post-Vietnam era. Perhaps the Soviet Union also has been more active in hopes of participating in East Asia's dynamic economic future. The torpid Soviet economy could surely benefit from it. The United States' recommitment to the security of East Asia under the Reagan administration since 1981 is likewise a response to the unprecedented Soviet military buildup in the area, as well as an obvious vote of confidence in the region's near-continuous economic performance.

Given this superpower propensity toward involvement and mutual escalation in East Asia, there also appears to be a shift away from their traditional Europe-centric focus. As the global "Asian-Pacific Era" takes hold, Washington and Moscow acknowledge the growing importance of the East Asian "flank" in their global strategic plans, although it is the United States, not the USSR, which does massive trade with the area. As the most economically dynamic, resilient, and growth-oriented region, East Asia and the Western Pacific is a critical international subsystem whose future developments will clearly affect the security of the superpowers.

This said, there are important differences in the two superpowers' motives, and their techniques of conflict management and resolution, in East Asia. While the Soviets have "lost" China as an ally—a critical loss—they have gained military facilities in Southeast Asia, are enlarging their facilities in Northeast Asia, have troops in combat on China's western flank, and have begun a number of South Pacific overtures. Under Gorbachev, they have proposed a "new model" of Asian-Pacific security and diplomatic relations, one in which the USSR is, inevitably, trying to utilize the region's tensions and contradictions to its advantage. The United States, by contrast, may be said to have "gained" China as an *informal* ally in complicating Soviet outreach; U.S. military forces in the region fell to such low levels in the 1970s and early 1980s that it threw the United States into a position of strategic dependence on its allies and friends' forces.

Unlike the earlier Cold War era in Asia (approximately 1947–69), which became dominated by the United States, the new period of Asian-Pacific economic dynamism and conflict management is characterized by a more aggressive and better-positioned USSR, opposed by countries associated with rather than dominated by the United States. Moreover, whereas political ideology, anticommunism, and fears of Communist subversion or takeovers acted as catalysts for allied action during the Cold War era, today it is rather the desire to protect the burgeoning market-oriented and largely capitalist economic structure of the coalition that inspires Asian/Pacific countries to rally behind U.S. efforts.

I. Regional Complexity and Deescalation Opportunities

Given these altered motives and changing power balances in the midst of continuing U.S.-USSR rivalry, East Asia's conflict patterns in the late 1980s are becoming increasingly complex. Two obvious examples are the Indochina situation and the Korean peninsula's problems, both conflict zones reflecting multiple participants, interests, and dilemmas. The complexity in East Asia's conflicts, in turn, reflects the tendency for bilateral conflict issues to become multilateral in scope, and to oscillate back and forth between domestic changes and external pressures. The existing pattern of predominantly bilateral relationships is giving way to multilateral diplomacy and region-wide interactions among East Asian countries, and between these Asian countries and the two principal outside powers, the United States and the Soviet Union. Each conflict in the region also has become enmeshed into "systemic" factors, such as the geopolitical location of the countries, border disputes marked by irrendentist and ethnic claims, local hegemonic desires for control of the commons—such as the open and enclosed seas—the availability of increasingly lethal armaments, and historic enmities between neighboring countries. Add to this the "spillover" of U.S.-Soviet geostrategic rivalry and the burgeoning complexity of these conflicts becomes obvious.

Consider, for example, these critical developments and their impacts on the region's conflicts and problems:

- Chinese and Soviet competition to woo ASEAN and Japanese friendship and trade as the two Communist giants, in turn, put an end to their estrangement.
- U.S. and Japanese competition for markets in China and among ASEAN countries as Washington and Tokyo remain fundamentally unable to resolve their own trade imbalance.
- North Korea's continuing diplomatic balancing act between Moscow and Beijing, but its increasingly closer military ties with the Soviet Union (which may, in turn, be producing tensions within the Pyongyang government).
- Vietnamese maneuvering between Moscow and Beijing within Hanoi's tighter alliance with the Soviet Union, while Hanoi under new pragmatic leadership perpetuates, for the time being, its efforts to place all of Indochina under its control.
- Emerging Sino-Soviet reconciliation and its impact on the superpower triangle and other Asia-Pacific alignments.
- Increased prospects for a Japanese-Soviet normalization of relations in the late 1980s.

- Territorial and economic conflicts among ASEAN countries regarding areas of the South China Sea.
- U.S.-Mongolian normalization of diplomatic relations.

In brief, a large array of regional conflicts and peaceful interactions have become more multilateral in scope, and that complexity now includes economic and diplomatic characteristics.

If the region is to continue developing as a world powerhouse and center for expanding trade and industry into the twenty-first century, the United States and its coalition allies must approach the challenge through a series of common actions, including demonstrating political stability among its individual Asian-Pacific member countries, assuring peace and security between them, but also presenting a common front toward the Soviets and their allies. The risks and complications to such a grand strategy are self-evident:

- Local conflicts can spill over into larger, region-wide contests (i.e., the Kampuchean anarchy encouraging Vietnamese-Thai rivalry which, in turn, draws in the Soviets, the Chinese, and the United States).
- The United States and the USSR may be drawn into local or regional disputes against their own best interests (hence the danger to all the major powers of the continuing Korean arms race).
- U.S.-USSR tensions may escalate into regional confrontations (i.e., the need for arms control regimes in the Sea of Japan, on the Korean peninsula, and in the South China Sea).
- Military clashes and intra-regional conflicts undermine economic dynamism and trade opportunities (e.g., Soviet exploitation of the Philippine insurgency perpetuates Cold War antagonisms in Southeast Asia).

As the United States and its partners attempt to steer the processes of conflict reduction in East Asia, new mechanisms and procedures for dispute settlement and conflict resolution seem necessary, and the feasibility of enhanced collective security measures and techniques needs to be explored. From that perspective, the following kinds of dispute settlement formula should be considered:

- Encapsulating conflicts within existing, or narrowed, boundaries so as to prevent spillovers and escalation (a key example is the need to demilitarize, hopefully neutralize, Kampuchea).
- Deescalating latent conflicts through preemptive measures of tension reduction such as increased communication and mutual exchange visits by cultural, economic, and political representatives (i.e., the confidence-building measures between North and South Korea, and the desirability of better conflict reduction efforts within ASEAN about South China Sea claims).
- Deescalating manifest conflicts through direct hot-line communications and jointly arranged diplomatic teams. (In this regard, the two Koreas have much of the necessary communications and negotiating machinery in place.)
- Settling conflicts through bilateral face-to-face negotiations and bargaining, third-party mediations and/or arbitration, and, where appropriate, through multilateral (collective security) machinery.

Techniques and formulas are, of course, only instruments in the hands of negotiators. More critical are the substantive deescalation proposals put forward at the bargaining table. At the broadest level, future diplomatic initiatives in East Asia which the American associated coalition should consider sponsoring include:

Category A. Mutual force reductions by the major powers in the region—the United States, the USSR, and China—involving ground, naval, and air forces. Particularly relevant tension zones for these proposals include the Sea of Japan area, the Sino-Soviet border, and Indochina. China's reduction of its armed forces by 1 million men factors into this category.

Category B. Mutual arms control and disarmament measures entailing freezing, then reducing, deployments of nuclear and conventional weapons. Clearly the Korean peninsula and the Sino-Soviet border apply here.

CONCLUSION 211

Category C. Nuclear-free zones involving region-wide considerations, subregions, or particularly sensitive locations. The ASEAN area and the Korean peninsula are two candidate zones, as are certain South Pacific areas.

Category D. An Asian-Pacific regional security conference, such as the Helsinki European Security Conference. The objective would be agreement on the general principle of attaining a substantial demilitarization of the Asian-Pacific region. The July 1986 Gorbachev proposals are an appropriate starting point.

II. Specific Measures for Intra-Regional Deescalation

Let us now turn to the individual conflict zones addressed in this book and specific proposals.

Korea

As one of the most lethally armed and dangerous conflict zones in East Asia, the need for less tension and more stability on the Korean peninsula is obvious. Both North Korea and South Korea are garrison states, the recipients of some of the most deadly and expensive weapons in the world. Perpetual military preparations distort both countries' priorities. How can the Korean situation be guided toward more stability and less tension? Given the peninsula's penetration by numerous external interests and factors, almost every category of general diplomatic proposals presented earlier in this chapter has relevance to Korea. Drawing on Scenarios 3 and 4 from Professor Kihl's Korea chapter—reducing tensions and institutionalizing the peace process between Seoul and Pyongyang—we note these possible initiatives:

a. Discussions by Washington and Moscow aimed at restraining and then perhaps halting altogether further deliveries of advanced fighter aircraft and missiles to their two respective Korean clients.

b. Conventional arms reductions by the two Koreas. Given Pyongyang's forward advantage of having combat troops only 35 miles away from Seoul, North Korea should offer a unilateral force pullback in return for an appropriate response by South Korea, perhaps a thinning out of ROK forces north of Seoul.

c. Collaboration between the United States, China, and the

USSR on the possibility of establishing a nuclear-free zone in and around the Korean peninsula.

d. Acceleration of inter-Korean negotiations and bargaining with the objective of gaining a nonaggression pact or peace treaty.

The Sea of Japan

As the most dangerously armed body of water in East Asia, the home port of the Soviet Pacific Fleet, and a flashpoint for superpower naval and air confrontation, negotiating an arms control regime in the Sea of Japan is an obvious priority. Can the Soviets be induced into less threatening behavior in the area? That should be the primary object of allied policy. Leverage and incentives exist: Driven by the USSR's dismal economic performance, Secretary General Gorbachev has been signaling his desire for better relations with Japan in particular. As Professor Olsen writes: "Japan's potentials to seriously rearm and either devise a unilateral strategic posture or become a truly active partner of the United States are tremendous. Moreover, those potentials are clearly recognized by the USSR. . . ."

From Professor Olsen's and Professor Falkenheim's chapters, we can extract a valuable range of policy incentives and proposals. Basically, they come down to this: The Soviets could be offered a new economic and political deal in Northeast Asia—principally expanded trade with Japan and South Korea and accelerated negotiations on a peace treaty with Japan—*provided* they enter into arms regime and demilitarization negotiations involving the Sea of Japan and the Northern Territories (the latter question possibly being split off as a separate discussion). With Japan and South Korea brought into partnership with the United States, the three allies could propose to the Soviet Union a wide-ranging series of discussions on security and economic matters bearing on the Sea of Japan and the southern Kurile Islands. Points to be discussed would include:

Soviet actions

- An end to Soviet military harassment of Japanese sea and air space.

- A return to Japan, as a first step, of one half of the southern Kurile Islands—Shikotan and the Habomais chain.

Allied actions

- Renewed Japanese-Soviet economic talks, the objective being Soviet acquisition of Japanese technology in return for export to Japan of Soviet oil and gas.
- Accelerated discussions on a Japanese-Soviet peace treaty, and a pledge by Tokyo not to remilitarize returned territories.

Combined Soviet and Allied actions

- Draw downs of Soviet and U.S. naval tonnage in the Sea of Japan.
- Gradually increasing Japanese-Soviet commercial, cultural, and diplomatic contacts.

Should the Soviets balk, the Japanese should be encouraged to speed up their military modernization programs with an accelerated focus on closing off the Sea of Japan's three chokepoints in time of war. Indeed, an emerging *informal* trilateral security cooperation between the United States, Japan, and South Korea—something particularly troublesome to Moscow—could be impressed upon the Russians.

The Sino-Soviet Conflict

The conflict of the largest territorial scope and potentially most dangerous fallout on East Asia is the Sino-Soviet conflict. With roots going back for centuries and manifesting racial, territorial, ideological, and even leadership personality characteristics, the Chinese and the Russians have been working at their relations for over 400 years. The dangerous period in the late 1960s and early 1970s, when Soviet and Chinese troops exchanged fire on the Ussuri River border, was followed ten years later by the start of what became regular consultations between the two powers. Ten rounds of discussions at the vice-ministerial level were held in the years prior to 1987. Spurred by Brezhnev's initiatives, they have produced a restoration of economic, trade, tourist, and cultural links, much of it dormant for twenty years. Sino-Soviet trade may reach $14 billion over the five years between 1986 and 1990 as the Soviets seek to appeal to China's trading instincts.

Progress on resolving political and military questions has been much slower. The huge Soviet military buildup encircling China has resulted in vastly superior Soviet forces on the Chinese border, including thousands of Soviet tanks, IRBM and MRBM forces, and a huge array of tactical aircraft. In response, the Chinese have presented their "Three Obstacles" preconditions for improvement in Sino-Soviet relations. Secretary General Gorbachev's July 1986 Vladivostok speech may have broken the logjam. Not surprisingly, Chinese leader Deng Xiaoping reacted positively to the tone of Gorbachev's proposals, sensing an opportunity to loosen Moscow's grip on China's peripheries while also improving the overall relationship. Both countries could use a reprieve from mutual tension and military spending, although the Soviets are in more serious difficulties than the Chinese.

The United States and its allies and friends in East Asia and the Pacific are sensitive to the future of Sino-Soviet ties, but they can relax about the kind of Sino-Soviet detente that has occurred to date. It has been so gradual and deliberate as not to upset the overall power distribution in East Asia. To the extent the United States, in particular, has any influence on the process of tension reduction between Moscow and Beijing, its influence can perhaps be exercised through separate politico-military negotiations with the two parties. Specifically, U.S. arming of China must emphasize weapons and equipment which do not provoke the Soviets. Also by pressing the Soviets, in turn, not to transfer SS-20s or other weapons from Europe to Asia, the U.S. can contribute to tension reduction in East Asia. As for negotiating mechanisms, the Chinese and the Soviets established those in 1982, and no improvements or suggestions from other parties are needed. In short, by 1987 the deescalation of the Sino-Soviet conflict had been under way for almost eight years, and it augurs well for East Asia's future stability.

The Indochina Conflict

The Indochina conflict, like the Korean peninsula conflict, is a multilayered, complex confrontation. But the Indochina problem has longer, more intractable historic roots than the division of Korea. Accordingly, combat outcomes in corners of Kampuchea affect Thai-Vietnamese relations, and impact on Soviet, Chinese, ASEAN, and U.S. interests. Basically, the guerrilla warfare in Kampuchea reflects a proxy contest between Thailand and Vietnam—a rivalry, as Professor Turley's chapter points out, which is intense and durable, creating pressures for others to become involved.

From history's perspective, neither Thailand nor Vietnam is an innocent victim or bystander. And both countries' armed forces are, as a matter of fact, benefiting from the current warfare. Still, perpetual conflict in Indochina continues to bleed Vietnam's economy and also promotes an unfortunate military preponderance in Bangkok's priorities. Thailand, for the present, is determined to forestall Vietnam's establishment of control over Cambodia and Laos, while Vietnam's rulers have seen Indochina as a single integrated security zone over which it must exercise ultimate control. How can peace be brought to Indochina?

The pro-Thai, anti-Vietnamese option would perpetuate current policy and conflict; i.e., Thailand and the Sihanouk-led Khmer guerrillas would continue to fight, supported by the United States, China, and ASEAN. China's pressure would continue on Vietnam. The long-term goal of this option would be grinding pressures which, along with Vietnam's dire economic straits, would produce a change of policy in Hanoi. The most desirable result would be a pullout of Vietnamese troops from Kampuchea and power sharing in Phnom Penh on terms favorable to Bangkok. The December 1986 changes in Hanoi's leadership may open the way for more pragmatic Vietnamese thinking along these lines.

Alternatively, the hard-line Vietnamese option would have ASEAN, the United States, and China accept the current Vietnamese version of Indochina's future—a consolidation of the three territories under Communist regimes subservient to Hanoi's wishes. Reversing this trend would be calculated as beyond the control of the United States and its friends, and could only be brought about by sharp increases of Chinese and Thai military pressure upon Vietnam.

An alternative to this "zero-sum game" is suggested by Professor Turley in his chapter:

> Over the long run American interests will be served best not by any particular outcome in Cambodia but by a Southeast Asia comprised of stable regimes, economies more advanced than China's, and nations at peace. Such a region would be one envisioned by ASEAN statesmen in which rules of order were made by Southeast Asians themselves. It is a vision out of reach so long as Thailand and Vietnam contend for influence in the countries between them.

Within Turley's broader perspective, is there a possibility that Kampuchea could be neutralized? While this option seldom has been discussed, and clearly would require more complex political-military arrangements than the other two options, a neutralized Kampuchea would both ease Thailand's perceived security threat from Vietnam, and reduce Vietnam's military burden outside its borders, something likely to appeal to Hanoi's new pragmatic leadership. Inducements to Hanoi's leadership to accept this option would include new ASEAN offers of technical and economic assistance, perhaps some territorial concessions in the South China Sea, and a relaxation of Sino-Soviet acrimony regarding Indochina. Since there are precedents for neutralization within Indochina, a neutral Kampuchea should not be ruled out. Laos was formally neutralized in 1962 and a tripartite government operated there until 1975.

The South China Sea

While almost all ASEAN members, and two of the Indochina states, have offshore claims in the South China Sea, these and other issues in the zone are, in Professor Weatherbee's view, "firmly embedded in broader political and strategic considerations" in the region. The stakes are substantial. Should quarreling between ASEAN states jeopardize the association's ability to present a common front to Hanoi and Moscow on other issues, serious opportunities could be lost for fashioning a more stable and less threatening environment in Southeast Asia and its "geopolitical lake," the South China Sea.

Chinese pressure is felt in the South China Sea. The PRC

claims *all* the more than 200 islands, reefs, and shoals, citing its "ancient" suzerainty and "indisputable" sovereignty, but Chinese forces occupy only the Paracels, leaving the Spratlys to other contenders. Vietnam and the Philippines have targeted the Spratlys with claims and naval forces. Malaysia, also, has recently injected forces into the region, desiring secure sealanes between Peninsula Malaysia and Sarawak and Sabah, and as a means of emphasizing its dispute with the Philippines over Sabah (a claim which the Aquino government in Manila has not renounced). Indonesia in turn, has claimed an Exclusive Economic Zone (EEZ), which carries its resource boundaries far into the South China Sea and well into Malaysia's claimed EEZ. Thailand also projects its EEZ out into the Gulf of Thailand and the northern end of the Straits of Malacca.

In turn, ASEAN and Indochina countries have overlapping and conflicting claims, Thailand quarreling with both Vietnam and Kampuchea, and Indonesia with Vietnam over the evidently oil-containing Natuna Islands in the South China Sea.

There are a variety of proposals which can reduce friction in the South China Sea conflict zone and possibly bring the area's forces under a degree of multilateral control. Among these proposals are:

(a) ASEAN, as a grouping, should consider declaring an Exclusive Economic Zone (EEZ) as a means of dampening intra-ASEAN disputes.

(b) Within this framework, make settlement of the Malaysian-Philippine dispute over Sabah a top priority since it, in particular, jeopardizes ASEAN's broader political-diplomatic agenda.

(c) Push for a common ASEAN negotiating position regarding Vietnam's withdrawal from Kampuchea. The emergence of Nguyen Van Linh and other pragmatists in Hanoi may give ASEAN a more tolerant audience among Vietnam's leadership—tolerant toward ASEAN offers of economic benefits (including concessions on South China Sea claims) in return for a neutralization of Kampuchea.

(d) Support ASEAN's general call for a Zone of Peace, Freedom, and Neutrality in Southeast Asia (ZOPFAN).

(e) Within a ZOPFAN framework, seek an arms control regime in the South China Sea in which U.S., Soviet, and Chinese weaponry and forces—in the Philippines, in Vietnam, and in the South China Sea—are brought under discussion.

The Philippines

The government of President Corazon Aquino and Vice President Salvador Laurel celebrated its first year in office in February 1987 amidst some of the most difficult continuing challenges faced by any government or society in East Asia. Inheriting twenty years of political and economic malaise, and a rising Communist guerrilla challenge, the Aquino government was battered from the left and the right. It had no choice but to try to tackle extraordinary challenges simultaneously: Rekindling a stricken economy; opening up negotiations with both Marxist and Muslim insurgents; revitalizing a demoralized armed forces; drafting a new constitution; reestablishing democracy; deflecting moves toward a coup d'état; and recomposing the security relationship with the United States.

While these extraordinary, and largely domestic, challenges preoccupied the Aquino government, the stakes in the Philippines for Southeast Asia as a whole are also critical: The Philippine Islands lie across the oil routes and sealanes between the Malacca Straits and the Tsushima Straits, and thus almost midway between Southeast and Northeast Asia. The Philippines also continues to host U.S. military facilities at Subic Bay and Clark Field just 800 miles east of Soviet military power at Cam Ranh Bay, thus flanking the South China Sea and bolstering the confrontation between ASEAN and Indochina.

As a result, stability in the Philippines is an essential sine qua non for a cohesive allied policy of deescalating tensions in the Southeast Asian area. Accordingly, U.S. and coalition policy toward the Philippines should emphasize these priorities:

(a) Back the Aquino-Laurel-Ramos power structure as the most representative, democratic hope for the Philippines.
(b) Support international and Philippine domestic efforts to revitalize the Philippine economy with an emphasis on privately owned and market-oriented activities. (In this regard, more Japanese and Asian Development Bank assistance should be pursued.)
(c) While continuing to offer amnesty and jobs to all insurgents who will lay down their arms, Manila should permit no NPA or MORO forces to have territorial authority or predominant functions of government.
(d) Seek ways of reducing the psychological and political pressure on the Philippines of hosting U.S. military forces in the area. Perhaps "ASEANizing" the bases is one option. Another may be transferring some of Subic Bay and Clark's functions to other Southeast Asian countries.
(e) Propose to the USSR draw downs in U.S. naval tonnage and aircraft in the Philippines in return for comparable responses by the Soviets in Vietnam.

III. Concluding Thoughts

Conflict zones in East Asia and the Western Pacific constitute fertile fields for politico-military proposals designed to deescalate and stabilize the region. The success of these proposals will, of course, be conditioned by the ability of the United States and its partners to more effectively blend their interests, to create common negotiating strategies, and to convince the Soviets and Moscow's allies that they can benefit from entering into these discussions. For example, the Reagan-Gorbachev dialogue during the Iceland summitry of October 11–12, 1986, is encouraging. The Americans and the Soviets discussed the possibility of reducing Soviet Asian SS-20s to 100, and the equivalent reduction of U.S. missiles in Asia, as a package deal related to an eventual removal of Euromissiles.

This book's evidence suggests that the seven conflict zones treated here provide extraordinary opportunities for creative and far-reaching thinking on restructuring East Asia's security needs and deescalating the region's conflicts. The challenge for the future generation of leadership working on East Asia's problems is to steer the twin processes of economic growth and tension reduction so that broadening peace and prosperity can be shared by all Asian-Pacific countries that want it. As Rome was not built in a day, so a new structure of peace and security in East Asia and the Pacific will require years of painstaking effort. It is a difficult objective, but it is also a noble and inspiring objective, and by turning more of East Asia's "swords into plowshares," it can be realized as the global Asian-Pacific Era unfolds.

SELECT BIBLIOGRAPHY

The following is a select bibliography of the published materials on East Asian conflict zones arranged under the two general headings of (1) global and region-wide focus and (2) specific conflict zones and/or individual country problems. Only titles of those books written in English, which are readily available in the United States, are included as a rule; many of the numerous journal articles and monographs are excluded from the following list. Under each of the eight subheadings, an average of one dozen items are included. Most of the items listed are up-to-date, at the time of this publication in 1987, in the sense that they cover the preceding five-year period of 1981–86.

I. Global and Region-Wide Focus

I-a. Asia-Pacific Region in General

Benjamin, Roger, and Robert T. Kudrle, eds. *The Industrial Future of the Pacific Basin.* Boulder, Colo.: Westview Press, 1984.

Brezezinski, Zbigniew. *Game Plan: A Geostrategic Framework for the Conduct of the U.S.-Soviet Contest.* Boston: Atlantic Monthly Press, 1986.

Hoffheinz, Roy, Jr., and Kent E. Calder. *The Eastasia Edge.* New York: Basic Books, 1982.

Linder, Staffan B. *The Pacific Century: Economic and Political Consequences of Asian-Pacific Dynamism.* Stanford: Stanford University Press, 1986.

Morley, James W., ed. *Pacific Basin: New Challenges to the U.S.* New York: Academy of Political Science, 1986.

Nishihara, Masashi. *East Asian Security and the Trilateral Countries.* New York: New York University Press, 1986.

Rostow, W. W. *The U.S. and the Regional Organization of Asia and the Pacific, 1965–1985*. Austin: University of Texas Press, 1986.

Scalapino, Robert A., Seizaburo Sato, and Jusuf Wanandi, eds. "Asian Political Institutionalization." *Research Papers and Policy Studies*, No. 15. Berkeley: University of California Institute of East Asian Studies, 1986.

I-b. U.S.-Soviet Competition

Bok, Tan Eng Georges. *The USSR in East Asia: The Changing Soviet Positions and Implications for the West*. Paris: The Atlantic Institute for International Affairs, 1986.

Fukuyama, Y. F. *Military Aspects of the U.S.-Soviet Competition in the Third World*. Santa Monica: Rand Corporation, November 1983.

Hsiung, James, ed. *U.S.-Asian Relations: The National Security Paradox*. New York: Praeger, 1983.

Luttwak, Edward N. *The Grand Strategy of the Soviet Union*. New York: St. Martin's Press, 1983.

Segal, Gerald. *The Soviet Union in East Asia: Predicaments of Power*. Boulder, Colo.: Westview Press, 1983.

Solomon, Richard H., and Masataka Kosaka, eds. *The Soviet Far East Military Buildup: Nuclear Dilemmas and Asian Security*. Dover, Mass.: Auburn House Publishing Company, 1986.

Stephan, John J., and V. P. Chicchkanov, eds. *Soviet-American Horizons on the Pacific*. Honolulu: University of Hawaii Press, 1986.

Stuart, Douglas T., and William Tow, eds. *China, the Soviet Union, and the West: Strategic and Political Dimensions in the 1980s*. Boulder, Colo.: Westview Press, 1982.

U.S. House of Representatives, Committee on Foreign Affairs. *The Soviet Role in Asia: Hearings*. 98th Congress, 1st session, 1983.

———, Special Subcommittee on U.S.-Pacific Rim Trade, Committee on Energy and Commerce. *Hearings, June 18, 1985. The Soviet Role in Pacific Rim Trade: U.S.-Soviet Environmental Cooperation*. Washington, D.C., 1985.

Whelan, Joseph G. *The Soviets in Asia: An Expanding Presence*. Washington, D.C.: Congressional Research Service, March 27, 1984.

Zagoria, Donald S., ed. *Soviet Policy in East Asia*. New Haven, Conn.: Yale University Press, 1982.

I-c. Security, Economic, and Other Functional Issues

Buckingham, William, ed. *Defense Planning in the 1990's*. Washington, D.C.: National Defense University Press, 1984.

SELECT BIBLIOGRAPHY 223

Buss, Claude A., ed. *National Security Interests in the Pacific Basin*. Stanford: Hoover Press, 1985.

Kihl, Young Whan, and Lawrence E. Grinter, eds. *Asian-Pacific Security: Emerging Challenges and Responses*. Boulder, Colo.: Lynne Rienner Publishers, Inc., 1986.

McMillen, Donald, ed. *Asian Perspectives on International Security*. London: MacMillan, 1984.

Morrison, Charles Ed., ed. *Threats to Security in East Asia-Pacific: National and Regional Perspectives*. Lexington, Mass.: D. C. Heath & Co., 1983.

O'Neill, Robert, ed. *Security in East Asia*. New York: St. Martin's Press for the Adelphi Library, 1984.

Scalapino, Robert A., Richard L. Sneider, and Masataka Kosaka, eds. *Opportunities and Constraints on Asian Regional Cooperation*. Dover, Mass.: Auburn House Publishing Co., forthcoming.

Scalapino, Robert A., Seizaburo Sato, and Jusuf Wanandi, eds. *Asian Economic Development—Present and Future*. Research Papers and Policy Studies No. 14. Berkeley: University of California Institute of East Asian Studies, 1985.

———, eds. *International and External Security Issues in Asia*. Research Papers and Policy Studies No. 16. Berkeley: University of California Institute of East Asian Studies, 1986.

Stuart, Douglas. *Security in the Pacific Rim*. Brookfield, Vt.: Tower Publishing Co., 1986.

Thomas, Raju G. C., ed. *The Great-Power Triangle and Asian Security*. Boston: Lexington Books, 1983.

Tow, William, and William Feeney, eds. *U.S. Foreign Policy and Asia-Pacific Security: A Transregional Approach*. Boulder, Colo.: Westview Press, 1982.

U.S. Department of Defense, *Soviet Military Power*. Washington, D.C.: GPO, April 1984.

II. Specific Conflict Zones and Individual Country Problems

II-a. Japan

Barnett, Robert W. *Beyond War: Japan's Concept of Comprehensive National Security*. Washington, D.C.: Pergamon-Brassey's, 1984.

Chapman, J. W. M. *Japan's Quest for Comprehensive Security: Defense, Diplomacy, Dependence*. London: F. Pinter, 1983.

Lee, Chong-Sik. *Japan and Korea: The Political Dimension*. Stanford: Hoover Press, 1985.
McIntosh, Malcolm. *Japan Re-Armed*. London: Frances Pinter Publishers, 1986.
Olsen, Edward. *U.S. Japan Strategic Reciprocity*. Stanford: Hoover Press, 1985.
Rees, David. *The Soviet Seizure of the Kuriles*. New York: Praeger Publishers, 1985.
Satoh, Yukio. *The Evolution of Japanese Security Policy*. London: International Institute for Strategic Studies, 1982.
Taylor, Robert. *The Sino-Japanese Axis: A New Force in Asia?* New York: St. Martin's Press, 1985.
Tsurutani, Taketasugu. *Japanese Policy and East Asian Security*. New York: Praeger, 1981.

II-a. China

Chang, Pao-min. *The Sino-Vietnamese Territorial Dispute*. New York: Praeger Publishers, 1986. Also, for the Georgetown University Center for the Strategic and International Studies, Washington, D.C., 1986.
Ellison, Herbert J. *The Sino-Soviet Conflict: A Global Perspective*. Seattle: University of Washington Press, 1982.
Huan, Guo-Cang. *Sino-Soviet Relations to the Year 2000: Implications for U.S. Interests*. Washington, D.C.: The Atlantic Council of the U.S., 1986.
Lasaster, Martin L. *The Taiwan Issue in Sino-American Strategic Relations*. Boulder, Colo.: Westview Press, 1984.
Lawson, Eugene K. *The Sino-Vietnamese Conflict*. New York: Praeger Publishers, 1984.
Lovejoy, Jr., Charles and Bruce W. Watson, eds. *China's Military Reforms: International and Domestic Implications*. Boulder, Colo.: Westview Press, 1986.
Medvedev, Roy. *China and the Superpowers*. New York: Basil Plackwell, 1986.
Quested, R. K. I. *Sino-Russian Relations: A Short History*. Boston: G. Alden and Unwin, 1984.
Rozman, Gilbert. *A Mirror for Socialism: Soviet Criticisms of China*. Princeton: Princeton University Press, 1985.
Segal, Gerald. *Sino-Soviet Relations after Mao*. Adelphi Papers. London: International Institute for Strategic Studies, 1985.
Sino-Soviet Conflict: A Historical Bibliography. Santa Barbara: ABC-CLIO Information Services, 1985.

Tsui, Tien-hua. *The Sino-Soviet Border Dispute in the 1970s*. Oakville, Ontario: Mosaic Press, 1984.

II-c. The Two Koreas

Cumings, Bruce. *The Two Koreas*. Headline Series No. 269. New York: Foreign Policy Association, 1984.

Gelman, Harry and Norman D. Levin. *The Future of Soviet–North Korean Relations*. Santa Monica: Rand, October 1984.

Kihl, Young Whan. *Politics and Policies in Divided Korea: Regimes in Contest*. Boulder, Colo.: Westview Press, 1984.

Koh, Byung Chul. *The Foreign Policy Systems of North and South Korea*. Berkeley: University of California Press, 1984.

Plomomka, Peter. *The Two Koreas: Catalyst for Conflict in East Asia?* Adelphi Papers. London: International Institute of Strategic Studies, 1986.

Rhee, Sang Woo. *Security and Unification of Korea*. Seoul: Sogang University Press, 1984.

Scalapino, Robert A. and Jun-Yop Kim, eds. *North Korea Today: Strategic and Domestic Issues*. Berkeley: University of California Institute of East Asian Studies, 1983.

——— and Hongkoo Lee, eds. *North Korea in a Regional and Global Context*. Berkeley: University of California Institute of East Asian Studies, 1985.

Sneider, Richard L. *The Political and Social Capabilities of North and South Korea for the Long-Term Military Competition*. Santa Monica: Rand, January 1985.

Wolf, Jr., C., D. P. Henry, K. C. Yeh, J. H. Hayes, J. Schank, and R. L. Sneider. *The Changing Balance: South and North Korean Capabilities for Long-Term Military Competition*. Santa Monica: Rand, December 1985.

II-d. Vietnam/Indochina

Albin, David A., and M. Hood, eds. *Cambodian Agony*. Armonk, N.Y.: M. E. Sharpe, 1987.

Chandler, David P., and Ben Kiernan, eds. *Revolution and Its Aftermath in Kampuchea: Eight Essays*. New Haven: Yale University Southeast Asia Studies, Monograph Series No. 25, 1983.

Chang, Pao-Min. *Kampuchea Between China and Vietnam*. Singapore: Singapore University Press, 1985.

Clutterbuck, Richard. *Conflict and Violence in Singapore and Malaysia, 1945–1983*. Boulder, Colo.: Westview Press, 1985.

Dommen, Arthur Jr. *Laos: Keystone of Indochina.* Boulder, Colo.: Westview Press, 1985.

Elliott, David W. P., ed. *The Third Indochina Conflict.* Boulder, Colo.: Westview Press, 1981.

Etcheson, Craig. *The Rise and Demise of Democratic Kampuchea.* Boulder, Colo.: Westview Press, 1984.

Rosenberger, Leif. *Soviet Union and Vietnam: An Uneasy Alliance.* Boulder, Colo.: Westview Press, 1986.

Turley, William S., ed. *Confrontation or Coexistence: The Future of ASEAN-Vietnam Relations.* Bangkok: Institute of Security and International Studies, Chulalongkorn University, 1985.

Vickery, Michael. *Kampuchea: Politics, Economics and Society.* London: Francis Pinter Publishers, 1986.

II-b. The ASEAN and Its Members

Alagappa, Muthiah. *The National Security of Developing States: Lessons from Thailand.* Dover, Mass.: Auburn House Publishing Co., 1987.

Buszynski, Leszek. *Soviet Foreign Policy and Southeast Asia: Determinants and Regional Responses.* London: Croom Helm, 1986.

Crouch, Harold. *Economic Change, Social Structure and the Political System in Southeast Asia: Philippine Development Compared with the Other ASEAN Countries.* Brookfield, Vt.: Dover Publishing Co., 1986.

Jackson, Karl D., and M. Hadi Soesastro, eds. *ASEAN Security and Economic Development.* Research Papers and Policy Studies No. 11. Berkeley, Ca.: University of California, Berkeley, Institute of East Asian Studies, 1984.

Lim, Joo-Jock. *Territorial Power Domains, Southeast Asia, and China.* Singapore: Institute of Southeast Asian Studies, 1984.

Millar, T. B., ed. *International Security in the Southeast Asian and Southwestern Pacific Region.* New York: University of Queensland Press, 1984.

Pauker, Guy. *Government Responses to Armed Insurgency in Southeast Asia: A Comparative Examination of Failures and Successes and Their Likely Implications for the Future.* Santa Monica: Rand Corporation, June 1985.

Snitwongse, Kusuma, and Sukhumbahand Paribatra, eds. *The Invisible Nexus: Energy and ASEAN's Security.* Singapore: Executive Publications Ltd, 1985.

Tangsubkul, Phiphat, *ASEAN and the Law of the Sea.* Singapore: Institute of Southeast Asian Studies, 1982.

Tilman, Robert. *Southeast Asia and the Enemy Beyond: ASEAN Perceptions of External Threats.* Boulder, Col.: Westview Press, 1986.

U.S. Congressional Research Service. "Insurgency and Counterinsurgency in the Philippines." Report Prepared for the Committee on Foreign Relations, U.S. Senate. Washington, D.C.: U.S. Library of Congress, 1985.

Weatherbee, Donald E., ed. *Southeast Asia Divided: The ASEAN-Indochina Crisis*. Boulder, Colo.: Westview Press, 1985.

II-f. The Philippines

Poole, Fred and Max Vanzi. *Revolution in the Philippines*. New York: McGraw Hill, 1984.

U.S. House of Representatives, Committee on Foreign Affairs. *Assessing America's Options in the Philippines*. Washington, D.C.: US GPO, 1986.

U.S. Senate, Committee on Foreign Relations. *The Situation in the Philippines*. Washington, D.C.: US GPO, 1984.

———, Select Committee on Intelligence. *The Philippines: A Situation Report*. Washington, D.C.: US GPO, November 1, 1985.

Index

Abe-Shevardnadze Joint Communiqué, 60
Afghanistan, 8, 18; Soviet intervention in, 6, 7, 31, 39, 41, 43, 45, 56, 57–58; Soviet withdrawal (proposed), 8, 9, 12, 43
Aliyev, G., 78
Alliances: East Asia and Western Pacific, 13–17; *see also* Asian-Pacific alliance system
Amboyna Key, 129, 138
Amur (Heilongjiang) River Basin, 9, 35, 43
Anambas, 133, 137
Andaman, Sea, 135
Ang Bayan (newspaper), 183–84, 186
Annamite Cordillera, 166
Antonov, Alexsey, 80
ANZUS, 14, 17
Aquino, Benigno, 186, 195
Aquino, Corazon, 17, 23, 177, 178, 179, 184; CPP reaction and response to, 189–94; policy toward CPP/NPA, 195–96
Aquino government, 200–1, 217, 218–19; challenges and policy options, 194–98; leftists in, 194, 196–97; prospects for, 201–2; and South China Sea conflict, 128, 131
Archipelagic principle, 132–34, 140
Arkhipov, Ivan V., 33, 34, 35
Armed Forces of the Philippines (AFP), 180, 190, 191, 193, 194, 195–96, 197, 198, 202
Armistice Agreement (Korean War), 114, 119
Armitage, Richard, 191–92
Arms control, 118, 209, 210, 218
Arms race, 92–93; Korean peninsula, 105, 112, 114, 116, 118, 209; South China Sea, 124–25; Thai-Vietnamese, 168
Arms reduction, 211, 219
Arroyo, Joker, 194
ASEAN (Association of South East Asian Nations), 2, 3, 7, 11, 14, 17, 19, 208, 210; and Asian-Pacific alliance, 16–17; Chinese relations with, 181; conflicts in,

ASEAN *(cont.)*
209, 210; defense role of, 89; economic growth, 3–4; in Indochina conflict, 10, 149, 150, 151, 157, 160, 161, 164, 165, 166, 169, 170, 215–16; Kuala Lumpur summit, 131; in South China Sea conflict, 23, 24, 123, 124, 125, 128, 131, 134, 135–37, 138–41, 216–18; Thailand's role in, 165; and Vietnam, 5, 22; Working Group on ZOPFAN, 139–40
Asian Games (Seoul), 16, 102
Asian Development Bank, 199, 219
Asian-Pacific Era, 206–7
Asian states: and Soviet strategic concerns, 72–73
Asian-Pacific alliance system, 13–17; burden sharing/power sharing in, 90–94; and deescalation opportunities in East Asia, 209–11, 214, 218–19; strategy of, 17–25
Asian-Pacific security conference (proposed), 8, 10, 211
Australia, 2, 5, 15, 26n20; in Asian-Pacific alliance, 17; defense role of, 89
Austria, 51

Baikal-Amur Mainline (BAM) railroad, 63
Baker Plan, 198
Balabac Strait, 123
Balance of power, 5, 73, 89, 92, 151, 207; South China Sea, 125, 141
Barter, 34, 103
Bashi Channel, 123
BAYAN (front organization), 190
Baylosis, Rafael, 192
Belgium, 185
Berjaya Party (Sabah), 130
Black Current (*Kuroshio*), 74
Black market, 162, 163, 185
Blas Ople (labor minister, Philippines), 197, 198
Bonin Islands, 52–53
Border disputes, 208; *see also* Sino-Soviet border

INDEX

Boy Morales, Horatio, 183, 194, 195
Brazil, 200
"Breakout strategy" (Soviet Union), 11, 12
Brevie Line, 136
Brezhnev, Leonid, 10, 31–32, 37, 53, 54, 213
Brzezinski, Zbigniew, 98
Brunei, 2, 10, 26n20, 130, 135; in South China Sea conflict, 123
Brunei Bay, 130
Buffer states: Thailand, 155, 157, 165, 170
Buffer zone of influence, 75, 76, 84, 89
Burma, 2, 104
Burman (race), 155
Buscayno, Bernabe (alias Commander Dante), 181, 195, 198

Cairo Declaration, 50
Cam Ranh Bay, 2, 6, 7, 87, 157, 169, 218
Cambodia, 10, 41, 136, 215; Asian-Pacific alliance strategy toward, 22–23; in Indochina conflict, 150–51, 158, 159, 160, 161, 162, 163, 164, 165, 166, 169–70; possible outcomes in, 165; "stable war" in, 168; in Thai-Vietnamese rivalry, 152, 154, 155, 156; Vietnamese occupation of, 13, 39, 41, 43, 45, 149; *see also* Kampuchea
Cambodian coalition, 22
Cape Nosappu, 58
Caribbean Basin Initiative, 200
CBS Television, 29; "60 Minutes," 13
Cease-fire (proposed: Philippines), 192–94, 197
Central America, 185
Chaovalit Yongchaiyuth, 157, 162
Chernenko, Konstantin, 36, 37
China. *See* People's Republic of China (PRC)
Chinese Communist Party (CCP): relations with CPSU, 38–40, 43–44
Chokepoints (Sea of Japan), 76, 78, 85, 89, 90, 98, 213
Chun Doo Hwan, 103, 104, 111, 117
Civic action program (Philippines), 196, 200
Clark Air Base (Philippines), 2, 6, 177, 180, 188, 218, 219
Climate, 82; and Soviet security concerns, 73–74
Coalition government (proposed): Philippines, 193, 197
Coalition Government for Democratic Kampuchea (CGDK), 22–23, 150, 161
"Cobra Gold" (military exercise), 138

Cocom restrictions (hi-tech exports), 63
Cold War, 79, 98, 207, 209
Colonialism, 142, 153, 154, 155
Commission on Human Rights (Philippines), 197
Communications: conflict reductions through, 210
Communism, communists: fear of, 207; in Indochina, 155; leadership of world, 39–40; in Philippines, 23, 177–205, 218, 219; Thai, 161, 164–65, 167; Vietnamese, 153, 154, 156
Communist parties: relations among, 38–40; splinter, 185
Communist Party of the Philippines (CPP), 177–79; and Aquino government, 189–98; funds from abroad, 185–86, 188–89; and proposals for stability, 198–202; reaction and
Communist Party of the Soviet Union (CPSU), 36; relations with CCP, 38–40, 43–44; relations with CPP, 179–80, 181, 182–83, 184, 187, 188
Competition: inter-Korean, 118–19; for markets, 208; in South China Sea, 123–25
Conflict management, 207
Conflict patterns: East Asia and Western Pacific, 1–28
Conflict reduction, 209–10
Conflict resolution, 30; mechanisms for, 209–10; North/South Korea, 116–18, 119; South China Sea, 138; structure for (ASEAN), 131; *see also* Tension reduction (Asian-Pacific)
Conflict zones: deescalation measures for, 211–19
Continental Shelf zones: South China Sea, 125, 129, 132, 134, 135, 136, 137, 139
Counterinsurgency: Philippines, 195–96
CPT, 162, 167
"Crony capitalism" 198, 201
Cultural elements: in Sino-Soviet relations, 36; in Thai-Vietnamese rivalry, 155
Cultural exchanges, 29, 187, 210
Cultural relations: Sino-Soviet, 35–36

Da Nang, 2
Danwan (reef), 129
"Declaration of Unity" (Philippines), 197
Deescalation (East Asia): allied strategy of, 17–25; Korean Peninsula, 97, 116–18,

Deescalation *(cont.)*
  119; measures for intra-regional, 211–19; opportunities for, 206–20
Deescalation proposals, 210–11
Defense capability: South Korean, 108–9, 115–16
Defense posture: North/South Korea, 108–10
Defense preparedness: North/South Korea, 105–6
Defense spending: fair levels of, 90; Japan, 15; North/South Korea, 105–6, 107, 112–13, 115
Demilitarization, 211
Democracy (Philippines), 178, 189, 190, 196, 201, 218, 219; "incomplete," 198
Democratic Soldiers (Thailand), 162
Deng Xiaoping, 12, 13, 23, 24, 29, 37, 43, 100, 102, 214
Detente, 8, 12, 14, 54, 77, 79, 181, 182; Sino-Soviet, 31–32; *see also* Deterrence plus detente (policy)
Deterrence (strategy), 12
Deterrence plus detente (policy), 12, 17–25, 115
Developing countries: debt, 198
Diplomacy: East Asian initiatives (proposed) 210–11; multilateral, 208; territorial, 152
Diplomatic relations: China-U.S., 31; multilateral, 209; in resolving Korean conflict, 117–18; Sino-Soviet, 13, 32, 38; Soviet-East Asian, 207; Soviet-Japanese, 10, 50; Soviet-Philippine, 182
Disarmament (proposed), 118, 210
Disinformation, 183, 187
Dispute settlement formulas, 210–11; *see also* Conflict resolution
DMZ (Korea), 2, 16, 98, 106, 108, 109, 115, 117, 119
Dominican Republic, 200
Dongsha Islands (Pratas), 125, 126
Dulles, John Foster, 51

East Asia, 1–6; allied strategy of stability and deescalation in, 17–25; conflict patterns in, 1–28, 207–11; economic dynamism of, 3–4, 7, 13, 19–20, 206, 207; geopolitical role of, 4–6; multilateral diplomacy in, 208; opportunities for deescalation of conflicts in, 206–20
Economic dynamism: East Asia/Western Pacific, 3–4, 7, 13, 19–20, 206, 207; Japan, 3, 80; South Korea, 3, 19–20,

Economic dynamism *(cont.)*
  106–7, 110, 111
Economic factors/issues: in Japanese-Soviet territorial disputes, 62–64; in North/South Korea conflict, 83, 99, 106–7, 112, 115; in regional disputes, 208–9; in relation between U.S. and allies, 90; in Sino-Soviet relations, 29, 32–35; in South China Sea conflict, 126, 132–38; in Soviet-South Korean relations, 77; in Thai-Vietnamese rivalry, 157, 161, 164; in Vietnam, 21–22
Economic zones (200-mile): Japan, 55, 64; USSR, 54–55, 58–69; *see also* Exclusive Economic Zone(s) (EEZ)
Economy: Philippines, 180–81, 188, 194, 195, 198–200, 201–2, 218, 219
Educational exchanges, 35–36
Eisaku Sato, 52–53
El Salvador, 185
Enrile, Juan Ponce, 177, 178, 190, 194, 195, 200, 201
Ethnic conflict, 1, 2–3
Etorofu (island), 49, 50, 51, 53, 54, 56–57; strategic importance of, 61–62, 66, 75
Eurasia, 73
Exclusive Economic Zone(s) (EEZ), 217; ASEAN (proposed), 217; South China Sea, 125, 128, 132, 133–34, 135, 136, 139
EXIMBANK, 199
Expansionism (Soviet Union), 4, 5, 6–13, 72–73, 75, 206; deterrents to, 14, 15, 17, 18–19
Exports, 34, 199–200; Cocom restrictions on, 63

Falkenheim, Peggy L., 47–69, 212
Finland, 51
Finlandization, 73
Firyubin, Nikolai P., 57
Fishing rights issue: South China Sea, 123, 132, 134, 135, 136–37; in USSR/Japanese relations, 54–55, 58–59, 62, 64, 79–80
Five Power Defense Arrangement, 137–38
Flat (island), 128
Formosa Straits, 123
Fourth Korean Workers' Party Congress, 105–6
France, 64, 127; in Indochina, 152–53, 154, 155

Freedom of the seas, 72, 75–76, 133
Front organizations (Philippines), 182, 190–91, 197
Fukuda (foreign minister, Japan), 62
Future scenarios: North/South Korea, 112–16

"Gang of Five," 3–4
Gasper, Carlos, 184
"Geographical Pivot of History" thesis, 4
Geography: in North/South Korea military postures, 108; in Soviet security concerns, 73–74
Geopolitical location, 208; Korean peninsula, 98–99, 116; South China Sea, 123, 124, 125
Geopolitics: East Asia's role in, 4–6
Gorbachev, Mikhail S., 4, 7–8, 36, 63, 77–78, 207, 211, 212, 219; Asia policy, 8–11, 29, 143; foreign policy tactics of, 47; and North Korean relations, 101, 111; and Sino-Soviet relations, 36–37, 41, 42–43; and Soviet-Japanese relations, 48, 59, 61, 65–66, 79; Vladivostok speech, 6, 7, 8–9, 12–13, 18, 24, 65, 72, 76, 100, 142, 214
Gorshkov, Admiral, 88
Great Britain, 135; confrontation in South China Sea, 124, 125
Great powers, 5; and Indochina conflict, 149–50, 151, 157, 159, 160, 164, 169; in South China Sea, 138, 141–43; strategic mobility, 140; *see also* Major powers; Superpowers
Grinter, Lawrence E., 1–28, 206–20
GRIT (Gradual Reduction in Tension), 113–14; *see also* Tension reduction (Asian-Pacific)
Gromyko, Andrei, 7, 10, 48, 58, 59
Gross national product (GNP), 3–4, 19–20, 106, 107
Guerrilla warfare, 109, 193–94
Gulf of Thailand, 123, 134, 135, 136, 217
Gulf of Tonkin, 24, 123, 126, 135–36

Habomai archipelago (Habomais), 49, 50, 51–52, 53, 56, 58, 66, 75, 213
Hainan Island, 24
Harrison, Selig, 126
Hatoyama Ichiro, 50, 51
Hawaii, 26n20
Helsinki Accord, 10

Helsinki European Security Conference, 211
Heng Samrin regime, 11, 161, 166
High-level contacts: Japan-Soviet, 48, 57–58, 59–60; North/South Korea, 114n; Sino-American, 35; Sino-Soviet, 29, 32, 33, 37–38, 40
Hiroshima, 10
Hokkaido, 53, 55, 64, 74, 79–80, 82
Holland, 185
Honecker, Erich, 43–44
Hong Kong, 2n,3, 199
Honshu (island), 74, 82
Hu Na, 36
Huang Hua, 37
Huangyen (Scarborough Shoal), 125

Iceland summit meeting, 219
IISS (International Institute for Strategic Studies), 105
Indochina, 2, 3, 8, 10, 18, 19; cooperation within, 166; deescalation opportunities in, 207–8, 210, 214–16; potential unification, 165; "Principles on Relations Between Indochina and ASEAN" (statement), 139; in South China Sea conflict, 123, 138, 217; Soviet influence in, 11, 181; Thai-Vietnamese conflict in, 149–76; Vietnamese goals regarding, 21
Indonesia, 2, 10, 17, 22, 165; defense buildup, 137; in South China Sea conflict, 123, 132–34, 137, 138, 140, 141
Industry, 107, 209
International Conference on Nicaragua and for Peace in Central America, 185
International Conference on Peace and Security in East Asia and the Pacific, 187–88
International Monetary Fund, 199
International Olympics Committee, 103, 117n
Ireland, 185
Ishibashi Masashi, 79
Itu Aba (island), 127
Investment, 3–4
Izvestia, 56

Jakarta, 22
Jalandoni, Louis, 185
Japan, 1, 2, 199; in Asian-Pacific alliance, 15, 17; defense role of, 77, 81–82, 89, 90–93; economic growth, 3, 80; foreign policy, 50–51, 65; in Indochina

Japan *(cont.)*
conflict, 164; international role of, 5, 61; irredintist movement, 64; and Korean Peninsula, 83, 85–86, 97, 99; Mid-Term Defense Program, 15; pacifism, 11, 65, 79; and Philippines, 219; rearmament, 81–82, 92, 169, 213; relations with China, 56, 88; strategic concerns in Sea of Japan, 70, 71, 76, 79–80; surrender—World War II, 49, 50; territorial dispute in relations with Soviet Union, 10, 47–69, 75; *see also* Soviet-Japanese relations; U.S.-Japanese relations

Japan Military Review, 110
Japanese Defense Agency (JDA), 15, 56–57
Japanese Peace Treaty (1951), 125
Jaruzelski, General, 43–44
Johnson, Lyndon, 52–53

Kaigara Island, 58
Kakuei Tanaka, 10
KAL 007 shootdown, 56, 79
Kalayaan (Freedom) Islands, 127–28
Kamchatka Peninsula, 61–62, 71, 74
Kampuchea, 2, 11, 18, 24, 26n20, 136, 150, 209, 210, 215; guerrillas in, 14; in Indochina conflict, 150; neutralization of, 7, 216, 217; in South China Sea conflict, 123, 124, 136, 138–39, 217; Vietnam in, 12, 16, 21, 22, 127
Kampuchean People's National Liberation Front, 162
Kang Song San, 111
Kapitsa, M., 78
Katipunan (organization), 182
Kazakhstan, 9, 43
KBL, 196, 197–98, 201
Khieu Samphan, 22, 23
Khmer (people), 152, 155, 159, 160, 164, 165, 166
Khmer People's National Liberation Front (KPNLF), 22
Khmer Rouge, 22, 23, 149, 150–51, 158, 159, 160, 165, 215
Khrushchev, Nikita, 30, 38, 50–51
Kihl, Young Whan, 1–28, 97–122, 206–20
Kilusang Mayo Uno (The May First Movement—KMU), 182
Kim Dae-Jung, 16
Kim Il Sung, 83, 85, 86, 101, 103, 106, 107, 111, 117
Kim Jong Il, 19

Kim Young-Sam, 16
Kong Korm, 158
Kono Yohei, 56
Korea, 26n20, 199, 209; as buffer state, 75; reunification of, 98, 101, 112, 118
Korea Strait, 74, 82, 84, 98
Korean peninsula, 14; conflict in, 97–122; deescalation opportunities in, 207–8, 210, 211–12; as strategic fulcrum among major powers, 97, 99; *see also* North Korea; South Korea
Korean War, 98, 114, 119
Kosygin, Aleksei, 56
Kriangsak Chomanand, 160
Kuala Lumpur, 129, 131
Kukrit Pramot government (Thailand), 160
Kunashiri (island), 49, 50, 51, 53, 54, 56–57; strategic importance of, 61–62, 66, 75
Kunshiri Channel, 62
Kuomintang government, 125
Kurile Current (*Oyashio*), 82
Kurile Islands, 3, 8, 75, 80, 213; northern, 49, 51; southern, 11, 24, 49; Soviet-Japanese territorial dispute over, 48–50

Labuan (island), 129–30
Lange government (New Zealand), 17
Lao (people), 152, 155, 167
Laos, 2, 10, 26n20, 215; in Indochina conflict, 153, 154, 155, 156, 157, 158, 159, 160, 161, 165, 166, 167; neutralization of, 11, 216
Laurel, Salvador, 23, 195, 201, 218, 219
Law of the Sea, 123–24, 132, 136, 140
LDP. *See* Liberal Democratic Party (LDP) (Japan)
Leninism, 181
Levchenko, Stanislav, 184
Levine, Steven I., 29–46
Li Peng, 37, 39, 45
Li Xiannian, 37, 101
Liaowang (*Outlook*) (journal), 29
Liberal Democratic Party (LDP) (Japan), 27n24, 51, 52, 66
Liberation theology, 186
Limbang Salient, 135
Littoral states: South China Sea, 123–25, 126, 132
Long, Robert, 15

Mackinder, Halford, 4, 73
Magellan, Ferdinand, 145n15

INDEX 233

Mahan, Alfred, 4, 73
Major powers: cooperation in tension reduction, 211–12; in/and East Asia, 1, 2–3; force reductions in East Asia (proposed), 210; interests in Korean peninsula, 97, 98–102, 104, 116, 118, 119; *see also* Great powers; Superpowers
Makintosh, Malcolm, 94n6
Malacca Straits, 123, 134, 143n*1*, 217, 218
Malay (race), 155
Malaysia, 2, 10, 22, 26n*20*; Continental Shelf Act of 1967, 129; in South China Sea conflict, 23, 123, 129–32, 133, 134–35, 138, 140, 217; Tseng-mu Reef, 126
Malaysian Federation, 130
Mao Zedong, 14, 30, 109, 195
Maoism, 180, 181, 188
Marcos, Ferdinand, 17, 23, 130–31, 138, 178, 180, 182, 186, 195, 196, 197, 198, 199, 200; exile, 177; fall of, 189, 190, 191, 192
Marcos government (Philippines), 179, 181, 182, 201
Maritime space: rights/duties of users of, 123, 132–34; *see also* Freedom of the seas, law of the sea
Medvedev, Roy, 34
Mekong Basin, 156–57
Miangas (Palmas) island, 134
Military (the): Philippines, 177, 180, 194–95, 197, 200–1, 202; Thailand, 161–63, 165, 168; Vietnam, 163, 168
Military balance: great powers, 99; *see also* Balance of power
Military buildup: East Asia/Western Pacific, 6–7; *see also* Soviet Union, military buildup
Military conflict, 1, 2
Military relations: Sino-Soviet, 36–44
Military situation: Korean peninsula, 99, 101, 105–6, 108–10, 115–16, 117, 119; Thai-Vietnamese rivalry, 167–68
Mindanao, 134, 191
Miyazawa (foreign minister, Japan, 54)
Mochtar (foreign minister, Indonesia), 140
Mon (race), 155
Mongolia, 2, 10, 12, 25n*4*, 209; Soviet forces in, 39, 41, 43
Moro insurgency (Philippines), 131
Moro National Liberation Front (MMNLF), 131
Munro, Ross, 183

Nakasone (prime minister, Japan), 10, 15, 24, 65, 66, 81, 102
Nansha (Spratly) Islands, 125, 126, 127–29, 138
Nanshan (island), 128
National Alliance for Justice, Peace, and Freedom, 190
National Democratic Front (NDF) (Philippines), 179, 183, 185, 186, 187, 190, 194
National liberation movements, 181
Nationalist Youth (Philippines), 190–91
NATO, 91
Natuna Islands, 137, 138, 141, 217
Natuna Utara, 133
Netherlands Indies, 146n*38*
New Liberal Club (Japan), 56
New People's Army (NPA) (Philippines), 179–89, 190, 191–92, 193, 194, 195, 197, 198, 199, 200, 201–2, 219
New Zealand, 2, 17, 25n*4*, 26n*20*, 89, 140
Nguyen Co Thach, 158, 166
Nguyen Van Linh, 19, 21, 163, 217
Nicaragua, 188
Niskisch, Larry A., 193
Nixon, Richard, 10
Nonnuclear regimes, 140
North Korea (Democratic People's Republic of Korea [DPRK]), 2, 7, 21, 77, 82, 208, 210, 211–12; Asian-Pacific alliance strategy toward, 19–20, 24; conflict with South Korea, 83, 84, 97–122; dependence on Soviet Union, 12, 18, 25n*4*, 93, 101, 111, 115; economic potential of, 106–7; Four Great Military Policy Lines, 105–6; *juche* policy, 85; military posture of, 108–10; military strength of, 16, 105–6; navy, 85; political institutions, 110–12; and Seoul Summer Olympics, 103–4; as Soviet buffer state, 78; strategic concerns in Sea of Japan, 70, 71, 85–87; trends in, 100–1
Northeast Asia, 2, 5; Soviet military presence in, 11, 207, 212
Northern Territories, 80, 212; obstacles to/incentives for resolution of dispute regarding, 61–67; strategic significance of, 61–62, 64–65; in territorial dispute between Soviet Union and Japan, 47–69, 75
Northwest Pacific, 5
Norway, 185

Nuclear weapons, 1, 6–7, 17, 73; basing of, 72, 73; Korean peninsula, 108, 109, 112–13
Nuclear-free zone(s) (proposed), 118, 140–41, 211, 212

Ocampo, Stur, 193
Oil and natural gas reserves: South China Sea, 23, 123, 124, 126, 128, 134, 136, 137, 138, 142, 217
Oil shock (1973), 54
Oil supply/demand, 125
Okinawa, 51, 53, 86, 88
Olsen, Edward A., 70–96, 212
"100 minute war," 126
Ongpin, Jaime, 198

Pacific Basin, 3
Pacific Ocean: access to, 62, 99
Pacific Ocean disarmament conference (proposed), 10
Pacific Oceania, 2
Palawan (island), 128
Pan Asian Security Conference (proposed), 65
Panata (island), 128
Papua New Guinea, 2
Paracels, 24, 125, 126–27, 217
Paris Accords, 22
Park Chung Hee, 106
Parliamentarian talks: North/South Korea, 102, 117
Parti Bersatu Sabah, 130
Passport and visa issue (USSR/Japan), 54, 60
Peace Declaration (USSR/Japan), 51–52
Peace Offensive: Philippines, 192–94; USSR, 10, 65
Peace process: institutionalization of (Korean peninsula), 114, 116–18, 211
Peace zone (proposed): East Asia, 139–41; South China Sea, 142–43; *see also* ZOPFAN (Zone of Peace, Freedom, and Neutrality)
Peaceful settlement: structure for, 124; *see also* Conflict resolution
Peking Convention, 49
Pelaez, Emmanuel, 200
Peng Zhen, 33, 36–37
Peninsula Malaya, 133, 134, 217
"People Power" (Philippines), 178, 189
People's Army (Vietnam), 163
People's Liberation Army (PRC), 41
People's Republic of China, 1, 2, 3, 21, 25n4, 32, 52, 62, 199; in Asian-Pacific alliance, 14–15, 17; claims in South China Sea, 23–24; and Communist party of the Philippines, 180–81; defense role of, 5, 89; Dengist reform group in, 77; economic revitalization program of, 32–33; foreign policy, 31, 33, 42, 44; in Indochina conflict, 149, 150, 154, 156, 157–58, 160, 162, 163, 164, 167, 169–70, 215–16; and Korean Peninsula, 97, 99; as military power, 2, 14, 87; naval modernization, 124, 141; Open Door economic policy, 33; rapprochement with Japan, 53, 56; relations with ASEAN, 181; relations with North Korea, 86, 100, 101; relations with South Korea, 102; and Sea of Japan strategic concerns, 71, 87–88; and South China Sea conflict, 123, 125–27, 129, 135–36, 138, 141, 216–17; South Sea Fleet, 23; threefold conditions for normalization of relations with USSR, 12–13, 37, 38–39, 40–43, 44, 214 (*see also* Sino-Soviet relations); and Vietnam, 22; *see also* Sino-American relations; etc.
People's Republic of Kampuchea (PRK). *See* Kampuchea
Per capita income, 4, 20, 21
Peripheries: in/and Asian-Pacific alliance, 19–24
Petropavlovsk, 61–62, 70, 71, 74, 87
Phak Mai (political party, Thailand), 167
Pham Bao, 172n19
Phibunsongkhram (Thai field marshal), 154
Philippine Armed Force (AFP). *See* Armed Forces of the Philippines (AFP)
Philippine Sea, 24
Philippine Peace and Solidarity Council, 187
Philippines, 2, 3, 10, 18, 90, 125; Asian-Pacific alliance strategy toward, 23; Base Line act of 1968, 130–31; Communist threat in, 177–205, 209; constitution, 142, 218; deescalation measures for (proposed), 218–19; democratic transition in, 23, 178, 189, 190, 196, 198, 201, 218, 219; elections, 177, 178, 189, 192, 193, 196, 201; land reform, 194; political difficulties, 17, 142, 192–94, 201; proposals for stability in, 198–201; in South China

Philippines *(cont.)*
  Sea conflict, 23, 123, 127–28, 130–33, 134, 138, 217; strategic importance of, 218; U.S. bases in, 2, 11, 23, 124, 140, 141–42, 169, 177–78, 180, 181, 182, 188, 194, 197, 218, 219
Phipat Tangsubkul, 135
Phnom Penh, 22, 149, 156
Phnom Penh government, 21
Phu Quoc (island), 136
Phuoc Tuy Province, 127
Pichit Kullavanijaya, 165
PKP (Partido Komunista Ng Philipinas), 180, 181, 182, 187
PLO, 183
Pol Pot, 149, 158, 160
Poland, 58, 59
Policy options: for deescalation— North/South Korea, 112–18; U.S. toward Soviets in East Asia, 11–12
Political relations: Japanese-Soviet, 57–58, 59–60, 63
Political situation: North/South Korean, 99, 110–12, 119
Politics: East Asia, 1; global, 99; 124–25; Philippines, 17, 142, 192–94, 201; in relations between U.S. and allies, 90; in Sino-Soviet relations, 30–31, 34, 36–44; in South China Sea conflict, 124–25, 126, 132, 134, 142; in Thai-Vietnamese rivalry, 159–64
Polyansky, Dimitri, 56, 57
Population: East Asia, 4
Porkkala Naval Base, 50–51
Potsdam Declaration, 50
Power relationships: South China Sea conflict, 132
Power sharing: Asian-Pacific alliance, 90–94
Pravda, 34, 187–88
Prem Tinsulanond, 162, 175n.55
Productivity, global, 3
Protectionism, 34, 90
Puolo Wai (island), 136
Putiatin, Evgenii, 48–49
Pyongyang, 108

Racism, 36, 72
Rama VII, king of Siam, 152
Ramos, Fidel, 177, 178, 190, 191, 194, 195, 200, 201
Rand Corporation, 99, 110, 111
Rangoon bombing, 16, 20, 104

Reagan, Ronald, 120n4, 219
Reagan administration, 13–14, 194, 206
Red Cross talks (Korean peninsula conflict), 103, 117
Reed Bank, 128
Resource development, maritime, 132; *see also* Oil and natural gas reserves
Revolution: Indochina, 154–55; Philippines, 180–81, 184, 191–92, 193–94, 195–97, 201–2
"Rimland," 73
Rimland coalition politics, 5
Roman Catholic Church, 186
Rosenberger, Leif R., 177–205
Russo-Japanese War, 49, 79

Sabah, 124, 130–31, 134, 217
Saguisag (special counsel, Philippines), 194
Saiyud Kerdphol, 155, 175n54
Sakhalin Island, 48, 49, 50, 51, 60, 74, 80
Sakhalin-Kurile Island Exchange Treaty (1875), 49, 67n6
Salas, Rodolfo, 183, 184, 192, 201
Saleh, Datuk Harris, 130
San Francisco Peace Treaty, 49, 50
Sanchez (labor minister, Philippines), 194, 196
Sangihe islands, 134
Sarawak, 130, 135, 217
Sayaboury Province, 157
Sea of Japan (Nihonkai; aka Eastern Sea [Dong Hae]), 2, 6, 18, 19, 70–96, 209, 210; Asian-Pacific alliance strategy in, 24; deescalation measures (proposed), 212–13; Soviet access to, 73–74, 75–76, 82
Sea of Okhotsk, 61–62, 73–74, 75, 76, 82
"Sea Power Domination" hypothesis, 4
Sea space zones: South China Sea, 125; *see also* Maritime space
Seabed jurisdiction conflict: South China Sea, 123, 126
Sealanes, strategic, 127, 130, 141, 217, 218
Second Indochina War, 154, 156
Security interests, issues, 4; China, 40–41; collective, 209–10; in Japanese-Soviet relations, 57, 61–62, 64–65; in Korean peninsula conflict, 97, 100–1, 117; in Philippine Communist insurgency, 195–98; in Sea of Japan, 71, 82–87; in South China Sea conflict, 125; of Soviet Union, 65, 67, 72–78; in Thai-

Security interests, issues *(cont.)*
  Vietnamese relations, 156–59, 161–62, 164–68
Seoul (South Korea), 109
Seoul Summer Olympics, 16, 77, 84, 101, 103–4, 110–11, 117
Shevardnadze, Eduard, 7, 8, 11, 24, 38, 76, 111; visit to Tokyo, 48, 59–61, 65
Shikotan Island, 49, 51–52, 53, 56, 57, 66, 75, 213
Shunichi Matsumoto, 51
Siam. *See* Thailand
Siberia, 35, 49; economic resource development, 5, 18, 47, 53, 54, 63, 99
Sibulu Passage, 133
Sihanouk, Prince, 22, 23, 161
Sihanouk resistance coalition, 11
Silicon Valley, 3
Singapore, 2, 3, 4, 10, 17, 26n20; in South China Sea conflict, 123
Singapore Straits, 133
Sino-American relations, 11, 14, 30, 36, 42, 44–45, 87, 90, 141, 181, 207; ambassadorial talks, 31; and Indochinese conflict, 169; normalization of, 32, 35; Taiwan as issue in, 42, 44
Sino-Soviet alliance, 30, 45
Sino-Soviet border, 2, 6, 8, 18, 32, 39, 41–42, 43; deescalation proposals, 9–10, 12, 210
Sino-Soviet conflict, 99, 127; deescalation proposals, 213–14; origins of, 30–31, 39–40
Sino-Soviet Economic, Trade, Scientific, and Technical Cooperation Commission, 33, 34
Sino-Soviet relations, 6, 7, 8, 9–10, 11, 12, 15, 18, 29–46, 87, 100, 166, 207, 213–14, 216; in East Asian conflict patterns, 208; economic cooperation agreements, 9, 29; implications of, for U.S., 44–46; normalization of, 38–39, 40–44, 45, 46, 138
Sino-Soviet Treaty of Friendship and Alliance, 31
Sino-Thais, 158
Sino-Vietnamese border, 2
Sison, Jose Maria, 180–81, 184, 192, 193, 195, 197
Sitthi Sawetsila, 161
"60 Minutes," 13
SLOCs (Sea Lines of Communication), 82

Smirnov, Boris, 186–87
Sneider, Richard L., 115
"Solidarity" (labor front organization, Philippines), 182
Solidarity groups (committees), 184–85
Son Sann, 22, 23, 162
Songjiana River, 35
Sonoda (foreign minister, Japan), 56
South China Sea, 2, 3, 6, 18, 19, 168, 210, 216; access/egress, 123; Asian-Pacific alliance strategy in, 23–24; conflict in, 123–48, 209; deescalation measures for (proposed), 216–18; maritime boundaries of, 125; as potential zone of peace, 124; "semi-enclosed sea" status, 123–24
South Korea (Republic of Korea [ROK]), 2, 7, 16, 110, 210, 211–12; in Asian-Pacific alliance, 16, 17, 19; conflict with North Korea, 97–122; defense role of, 89, 90–91; economic growth, 3, 19–20, 106–7, 110, 111; Five-Year Force Modernization Plan, 106; Force Improvement Plans I and II, 106; Force Improvement Plan III, 115; GNP, 3; military posture, 108–10; military strength, 14, 105–6; political institutions, 110–12; Soviet Union as threat to security of, 83–85; strategic concerns of, regarding Sea of Japan, 70, 71, 76, 78, 82–85; trends in, 101–2
South Pacific, 18, 207, 211
South Pacific Forum: Nuclear Free Zone Treaty, 140
South Vietnam, 136
South Yemen, 183, 185
Southeast Asia, 2, 141, 169; economic growth, 3–4; races in, 155; Soviet influence in, 11, 207; U.S. policy in, 169–70
Soviet-American relations, 56; in Asia, 4–6, 10–11; balance of power in, 207; and complexity of conflict patterns in East Asia, 208–10; *see also* Superpowers
Soviet-Chinese relations. *See* Sino-Soviet relations
Soviet Committee for the Defense of Peace, 187
Soviet-Japan Friendship Association, 64
Soviet-Japanese relations, 8, 10, 18, 24, 25–4, 76–77, 79–81, 208; cooperative agreements in, 59, 61; and deescalation

Soviet-Japanese relations *(cont.)*
  of conflict, 212–13
Soviet Navy, 6; Baltic Sea Fleet, 74; Black Sea Fleet, 74; Far Eastern Fleet, 62; Northern Fleet, 74, 88; Pacific Fleet, 2, 41, 70, 71–75, 76, 84, 87, 88–89, 212
Soviet Ocean Shipping Company, 35
Soviet Union: Asian-Pacific alliance and, 5, 18–19; and Communist Party of the Philippines, 179–80, 181–89, 190; in/and East Asia/Western Pacific, 1, 2, 4–13, 24, 29, 206–7; expansionism, 4, 5, 6–13, 14, 15, 17, 18–19, 72–73, 75, 98, 206; goal of breaking up U.S.-Japanese-Chinese-ASEAN alliance, 8, 11, 24, 76, 81, 90, 91; in Indochina conflict, 149–50; and Korean Peninsula, 76, 77–78, 83–85, 93, 97, 99, 100, 102, 111, 115; and Marcos government, 189–90; military buildup in Asia-Pacific, 2, 40–41, 56–57, 61–62, 65, 67, 72, 79, 88, 99, 206, 207, 214; as Pacific power, 70, 71, 72–73, 141, 206; "peace offensive" foreign policy, 10, 65; policy in East Asia, 12, 206–7; proposed joint ventures, 10, 43; psychological insecurities, 72; security interests/military objectives of, 65, 67, 72–78, 180; in South China Sea conflict, 124, 125, 127, 129, 138, 140–43; strategic concerns in Sea of Japan, 70, 71–78, 86–87, 90–91, 212–13; strategic parity with U.S., 54, 61; support for Vietnam, 12, 18, 25n4, 157, 163; territorial dispute in relations with Japan, 47–69; trading partners, 3, 25n4; *see also* Afghanistan
Sovietskayagavan, 70, 71
Soya Strait, 74, 82
Sports talks: North/South Korea, 103–4
Spratly Island (Truong Sa), 127, 128–29, 217
Spratly Islands. *See* Nansha (Spratly) Islands
Spykman, Nicholas J., 73
Stability (East Asia), 1, 3; allied strategy of, 17–25
Stalin, Joseph, 38
Star Wars, 24
"Starfish" (military exercise), 138
Strategic concerns: East Asia, 14; in Japanese-Soviet relations, 64–67, 81; in North / South Korea military postures,

Strategic concerns *(cont.)*
  108–10; in Sea of Japan, 70–96; in South China Sea conflict, 126, 140
Strategic Defense Initiative (SDI), 8, 10, 81
Strategic environment: Korean Peninsula, 97, 98–104, 118
Strategic plans, global, 206–7
Strategies and capabilities: Thai-Vietnamese conflict, 154–55
Subic Bay, 2, 6, 177, 180, 188, 218, 219
Suharto (president, Indonesia), 17
Sulu Archipelago (Philippines), 131
Sulu Sea, 123, 133
Sunda Straits, 123
Superpowers, 83, 206–7, 219; balance of power in Asia, 92; China's stance toward, 42; competition, East Asia, 13, 98; conflict in Sea of Japan, 71; *see also* Great powers; Major powers; Soviet Union; United States
Sweden, 185
Systemic factors: in East Asian conflict, 208

Taiwan, 2, 3, 10, 27n23, 31, 32, 199; claims in South China Sea, 23, 24; economic growth, 3; as issue in Sino-American relations, 42, 44; in South China Sea conflict, 123, 126
Taiwan Straits, 10
Takashima Masuo, 57
Talaud islands, 134
Tanaka (prime minister, Japan), 53, 54
Tanaka-Brezhnev Joint Communiqué, 60
Tatar Strait, 74
"Team Spirit" (military exercise), 103, 117
Technology, 3, 63
Tension reduction: Asian-Pacific, 92, 93; Korean peninsula, 103, 113–14, 116–18, 211; Sino-Soviet relations, 31–32
Territorial baselines: South China Sea, 132–34
Territorial conflict: East Asia, 1, 2–3; South China Sea, 23–24, 123, 124, 125–32, 217; Soviet-Japanese: Northern Territories, 47–69, 75
Territorial diplomacy, 152
Terrorism, 7, 20, 104, 197
Terumbu Layang-Layang, 129, 137, 138
Thai (race), 155
Thailand, 2, 4, 10, 11, 14, 138, 181; in Asian-Pacific alliance strategy, 19;

Thailand *(cont.)*
  domestic politics, 17, 159–63; EEZ, 217; foreign policy, 160–61; in Indochina conflict, 150–51; Internal Security Operation Command, 175n55; military assistance pact with U.S., 154, 156; refugees in, 153; in South China Sea conflict, 123, 134, 135, 136–37
Thai-Vietnam rivalry: history of, 152–56; Indochina conflict, 149–76, 209, 215–16; perceptions of intentions in, 152, 156–59, 164; strategies and capabilities in, 164–68
Thalweg principle, 137
Thanat Khoman, 155
Third Indochina War, 149, 150–51, 156
Third World, 113
Thithu (Pagasa), 127, 128
To Huu, 21
Tolkunov, Lev N., 37–38
Tonkin Gulf. *See* Gulf of Tonkin
Trade, 3–4, 90, 209; issues in U.S./Japan, 54, 208; in Japanese-Soviet relations, 63, 80–81; Sino-American, 34; Sino-Soviet, 29, 32–35
Treaty of Portsmouth, 49
Treaty of Shimoda, 48–49
Tsugaru Strait, 74
Tsushima Straits, 82, 84, 218
Turley, William S., 149–76, 215, 216

United Nations, 50, 58, 161; General Assembly, 53, 139
United States: and Asian-Pacific alliance, 13–14, 17; defense role, 88–89; East Asia policy, 1, 2, 11, 98, 169–70, 206–7; foreign policy risks in Asian-Pacific alliance, 18–19; implications of ameliorization of Sino-Soviet relations for, 44–46, 214; in Indochina conflict, 149–50, 151, 157, 160, 161, 164, 168–70, 215–16; and Korean peninsula, 90, 97, 99, 100, 101; military assistance pact with Thailand, 154, 156; military bases in Philippines, 2, 11, 23, 124, 140, 141–41, 169, 177–78, 180, 181, 182, 188, 194, 197, 218, 219; military presence in Asian-Pacific area, 2, 14, 15, 26n20, 54, 180, 207; policies toward Soviets in East Asia, 11–12; political/military

United States *(cont.)*
  presence in South China Sea, 124, 125; role in Philippines, 23, 90, 177–78, 179, 180, 194, 197, 199–200; role in Soviet-Japanese relations, 51, 52–53, 65; security relations with Australia, 17; in South China Sea conflict, 128–29, 138, 140–43; and South Korean defense, 83, 84, 85–86, 90–94, 100, 101, 108–9, 110, 115; strategic concerns regarding Sea of Japan, 70, 71, 76, 78, 88–89; strategic parity with Soviet Union, 54, 61; strategic position in East Asia, 44; tensions between allies and, 90; trade with East Asia, 3, 4, 208; and Vietnam, 22; *see also* Sino-American relations; Soviet-American relations
U.S. Department of State, 114n
U.S.-Japanese relations, 14, 54, 90, 99, 208; in defense, 15, 65, 90–93
U.S.-Japanese Security Treaty, 52
U.S. Navy, 141; Seventh Fleet, 2, 6, 10, 85, 88, 138
U.S.-Philippines Mutual Defense Treaty, 128
Uomoto Tokichiro, 57
Uruppu (island), 49

Ver, General, 195, 200
Vientiane, 11, 152, 156
Vietnam, 2, 3, 7, 10, 11, 208; Asian-Pacific alliance strategy toward, 19, 21–22; claims in South China Sea, 23, 24, 217; dependence on Soviet Union, 21; domestic politics, 163–64; historical rivalry with Thailand, 152–56, 209; Hua Phan Ha Than Ghok region, 152; in Indochina conflict, 150–51; invasion/occupation of Kampuchea (Cambodia), 12, 13, 39, 41, 43, 45, 127, 160, 215; per capita income, 4; Sip Song Chau Thai region, 152; in South China Sea conflict, 123, 124, 126–27, 128, 129, 134, 135–37, 138; Soviet bases in, 16, 41, 124, 140, 141, 142, 157, 169; Soviet relations with, 12, 18, 25n4, 157, 163
Vietnam War, 181
Vietnamese-Thai rivalry: in Indochina conflict, 149–76; perceptions of intentions in, 152, 156–59
Virata, Cesar, 128

Vladivostok, 10, 70, 86, 87; opening to visitors by foreigners, 9; Soviet basing, 6, 7

Weatherbee, Donald E., 123–48, 216
West Germany, 52, 54, 64, 185
Western Pacific: conflict patterns in, 1–28; economic dynamism in, 3–4; overview of, 1–6
Wilhelm, Alfred, 42
World Bank, 199
World Federation of Trade Unions (WFTU), 182
World Peace Council (WPC), 187–88
World War II, 49, 50, 52, 79, 154, 158
Wu Xingtang, 38
Wu Xueqian, 38, 40, 43

Xinjiang Uighur Autonomous Regions of China, 9, 43

Xisha Islands. *See* Paracels

Yalta Agreement, 50
Yalta Conference, 49
Yanai Shinichi, 65
Yao Yilin, 29, 33, 35
Yellow Sea, 78, 87
"Young Turks" coup (Thailand), 160, 162

Zhao Ziyang, 34
Zheng Chengxian, 37
Zhongsha (Macclesfield Bank), 125, 126
Zhou Enlai, 38, 125
Zone of Peace, Freedom, and Neutrality (ZOPFAN), 139–41, 169
ZOPFAN (Zone of Peace, Freedom, and Neutrality), 169, 218